Handbook of Diagnostic Biochemistry and Hematology in Normal Pregnancy

Edited by

Gillian Lockitch, M.D., F.R.C.P.(C.)

Associate Professor
University of British Columbia
Program Head
British Columbia's Children's Hospital
Consultant Pathologist
Grace Hospital
Vancouver, British Columbia
Canada

CRC Press
Boca Raton Ann Arbor London Tokyo

Library of Congress Cataloging-in-Publication Data

Handbook of diagnostic biochemistry and hematology in normal pregnancy/
edited by Gillian Lockitch.
p. cm.
Includes bibliographical references and index.
ISBN 0-8493-3518-3
1. Pregnancy—Physiological aspects—Tables. 2. Blood—Analysis—
Tables. 3. Reference values (Medicine)—Tables. 4. Clinical
chemistry—Tables. 5. Hematology—Tables. I. Lockitch, Gillian.
[DNLM: 1. Pregnancy—blood—handbooks. 2. Pregnancy—physiology—
handbooks. 3. Reference Values—handbooks. WQ 39 H23555]
RG558.H36 1992
618.3'07561—dc20
DNLM/DLC
for Library of Congress 92-48712
CIP

Direct all inquiries to CRC Press, Inc., 2000 Corporate Blvd., N.W., Boca Raton, Florida 33431.

International Standard Book Number 0-8493-3518-3

Library of Congress Card Number 92-48712

Printed in the United States of America 2 3 4 5 6 7 8 9 0

Printed on acid-free paper

PREFACE

The need for this handbook was perceived approximately 3 years ago, when a review of our computer-generated reports of biochemistry test results from pregnant women indicated that a disproportionate number of results were flagged as abnormal. This clearly resulted from the fact that our system contained a single reference range for adult women for each analyte, derived from reference value studies in healthy non-pregnant women rather than pregnancy-specific reference ranges. Consultation with clinical colleagues, obstetricians, and obstetric anesthesiologists revealed that most ignored the ''out-of-range'' and ''panic'' flags, with the recognition that ''we know that these values are different in pregnant women''. However, no one had a clear concept of the stage at which these differences were significant for various analytes or just how different the expected values should be.

It is known that the concentrations of many biochemical and cellular elements in blood change markedly with increasing length of gestation. This results from physiological alterations in hormone production, increase in blood and plasma volume, and adjustments in systemic distribution of blood flow. Unless these gestation-specific changes in blood chemistry and hematology are incorporated into the reference ranges used in interpretation of laboratory tests, many results of blood tests from pregnant patients may appear to be significantly outside the reference range, and thus abnormal.

Few, if any, laboratories report results with a comprehensive system of pregnancy reference ranges derived from healthy women tested sequentially throughout pregnancy. Pregnancy-specific reference ranges are generally used for some tests, such as the reproductive hormones or for specific screening tests such as the glucose loading test for gestational diabetes or α-fetoprotein testing for neural tube defects. However, for most other tests, laboratories in general provide non-pregnant adult reference ranges against which physicians must interpret results for their pregnant patients.

One reason for this is probably lack of a single source providing a comprehensive set of data showing the sequential changes that occur in the commonly available biochemical and hematological tests as gestation progresses. With the encouragement of our obstetrical and anesthesiology colleagues we aimed to develop this source, using data from a single group of healthy well-nourished pregnant women followed in a prospective longitudinal study, from the first trimester to the 6th week postpartum visit. The material in this book is based on the laboratory handbook which we developed for physicians providing obstetrical care to women attending the Grace Hospital in Vancouver, British Columbia.

We show in graphic and tabular form pregnancy-specific variations for hematological and biochemical diagnostic tests *routinely* available from our laboratory. Included are data for over 70 analytes, including electrolytes, minerals, enzymes, proteins, hormones, and nutritionally related parameters such as trace elements and vitamins. Since absolute values should not be directly applied without verification to a test performed on different equipment or using different reagents, the data are presented graphically to show trends, as well as in tabular form showing variation from a non-pregnant control mean value.

The results from our studies are compared with studies reported in the literature. Any notable differences are discussed in the commentary which accompanies each section. Since the sample demographics, study design, specimen collection and handling protocols, and test methodology (instrumentation and specific reagents) all affect the final results, this information is provided in some detail for our data.

There are important caveats to be understood by users of this handbook. Our objective was to provide a framework of biochemical and hematological reference intervals in healthy well-nourished women living in an environment which provides ample health care and nutritional resources to promote healthy pregnancy.

The reference data provided in this book, for both the non-pregnant control women and the pregnant women, were derived from a very specific population group consisting of healthy, well-educated, middle class women with easy access to health care and adequate nutrition. Many were physicians, laboratory technologists, nurses, nutritionists, students, and other professionals. Most were Caucasian, with 6% of the pregnant subjects and 10% of the control subjects of Asian origin.

The analytical methods used for specimen testing are those in routine use in the Clinical Biochemistry and Hematopathology Diagnostic Laboratories serving the British Columbia Children's Hospital and the Grace Hospital, the tertiary care university affiliated pediatric and obstetric hospitals for the province of British Columbia. All methods are subject to stringent internal quality control and monitored by external proficiency testing programs. The specific test methods are described in the last section.

We emphasize that the absolute reference intervals should not be used directly to interpret results obtained by different methods without validation of their transferability. For those tests which exhibit varying performance characteristics with different reagents and instrumentation, comparisons with the data in this book should be made by comparing the mean from the non-pregnant sample shown in the tables with the results of similar tests in one's own laboratory. The "percentage of control mean" column can then be used to establish comparative values at various stages of pregnancy.

To assist in comparison of our values with those of other laboratories, we have provided a section showing reference intervals for the biochemical and hematological tests derived from healthy non-pregnant women. This will enable a comparison to be made with the reference intervals already in use in a given laboratory. Here it is also important to note that use of oral contraceptives can alter the concentrations of many analytes, in particular proteins, hormones, and coagulation factors. We provide reference intervals which both include and exclude oral contraceptive users. For analytes where genetic or racial population differences are recognized, the ethnic composition of our study sample must be taken into account.

The final pregnancy study sample, after exclusion of those women who failed to meet the criteria of uncomplicated pregnancy, delivery, and healthy singleton infant, consisted of 29 women. The provision of reference limits based on this sample size may be questioned by statisticians and epidemiologists. Practicing physicians and obstetricians will however appreciate the extreme motivation required on the part of the participants who provided blood specimens on nine occasions over the course of pregnancy. Pathologists will appreciate the magnitude of the analytical undertaking, which involved almost 20,000 individual tests, excluding quality control or confirmatory assays. We consider that the longitudinal design of the study in which each women serves as her own referent increases the value of the data. The non-pregnant control group of 121 healthy women from the same professional and social environment provided a sound base of reference ranges against which the pregnancy variation could be compared.

The objective of prenatal and perinatal care is the successful completion of a healthy uncomplicated pregnancy and puerperium, with the birth of an appropriately grown full-term healthy infant. The ability to differentiate definitively abnormal laboratory results from those physiologically altered by pregnancy should aid in the early recognition of pathological processes which may adversely affect outcome for mother or infant.

Our data provide an insight into the natural changes that occur in the blood biochemistry and hematology under optimal conditions of pregnancy in a healthy Western population receiving excellent antenatal care.

It is our intent that this handbook will provide useful practical information for physicians and other health care professionals caring for pregnant women, as well as for laboratories that analyze blood specimens for clinical and research purposes from this distinctive population.

THE EDITOR

Gillian Lockitch, M.D., F.R.C.P.(C.), is Head of Clinical Biochemistry at the British Columbia Children's Hospital and Associate Professor of Pathology in the Faculty of Medicine at the University of British Columbia, Vancouver, Canada.

Dr. Lockitch received her Bachelor of Medicine and Surgery degree from the University of Cape Town in 1968. From 1971 to 1974 she was a Senior Research Officer in the Division of Medical Statistics and Epidemiology in the Nutritional Research Institute of the South African Medical Research Council. During this time she was awarded the Boonzaaier Prize for research in 1972 and the Andries Blignaut Medal in 1973. In 1974 she earned the Doctor of Medicine degree from the University of Cape Town.

On completion of residency training at the University of British Columbia, Dr. Lockitch gained certification as a Fellow of the Royal College of Physicians of Canada in Pediatrics in 1978 and in Medical Biochemistry in 1980. She was appointed to her current position in 1982. The Clinical Biochemistry Program which she heads, provides diagnostic services to both the Children's Hospital and the Grace Hospital, the tertiary care university affiliated pediatric and obstetric hospitals for British Columbia.

Dr. Lockitch is a member of the Canadian Association of Medical Biochemists, the Canadian Academy of Clinical Biochemists, the Canadian Society of Clinical Chemists, the American Association for Clinical Chemistry, the International Society for Trace Element Research in Humans, and an associate member of the Society of Obstetricians and Gynecologists of Canada.

She is a member of the Royal College Specialty Committee for Medical Biochemistry, past-president of the British Columbia Society of Clinical Chemists and is currently the Chair of the Pediatric Clinical Chemistry Division of the American Association for Clinical Chemistry.

Dr. Lockitch has received research grants from the British Columbia Research Foundation, British Columbia Medical Services Foundation, and private industry. She has published 45 research papers and given more than 30 invited presentations at national and international meetings. Her major research interests include the biochemical and nutritional physiology of pregnancy, perinatology, and infancy, with particular emphasis on the importance of trace elements and ultratrace elements in maternal and child health.

CONTRIBUTORS

Anne Catherine Halstead, M.D., F.R.C.P.(C.)
Clinical Associate Professor
Department of Pathology
University of British Columbia
and Medical Biochemist
Division of Clinical Biochemistry
British Columbia's Children's Hospital
and Consultant Pathologist
Grace Hospital
Vancouver, British Columbia
Canada

Gillian Lockitch, M.D., F.R.C.P.(C.)
Associate Professor
Department of Pathology
University of British Columbia
and Program Head
Division of Clinical Biochemistry
British Columbia's Children's Hospital
and Consultant Pathologist
Grace Hospital
Vancouver, British Columbia
Canada

Hilary Vallance, M.D.
Clinical Instructor
Department of Pathology
University of British Columbia
and Medical Biochemistry Resident
Division of Clinical Biochemistry
British Columbia's Children's Hospital
Vancouver, British Columbia
Canada

Louis Wadsworth, M.D., F.R.C.P.(C.), F.R.C.Path.
Clinical Professor
Department of Pathology
University of British Columbia
and Program Head
Division of Hematopathology/Transfusion Medicine
British Columbia's Children's Hospital
and Consultant Pathologist
Grace Hospital
Vancouver, British Columbia
Canada

Bernd Wittmann, M.D., F.R.C.S.(C.)
Adjunct Professor
Department of Psychology
Simon Fraser University
and Consultant Pathologist
Obstetrics and Gynecology
Department of Ultrasound
Grace Hospital
Vancouver, British Columbia
Canada

ACKNOWLEDGMENTS

This work was completed with the assistance of many individuals. The specimens were collected, processed, and analyzed by the staff of the Accessioning, Clinical Biochemistry and Endocrinology, and Hematopathology Programs of the Department of Pathology at British Columbia Children's Hospital, under the supervision of Esme Leigh, Gisela Black, Margaret Tkachuk, and Rosemary Coyne. Many people provided extra assistance with accessioning, special analyses, or data retrieval. Staff in the Grace Outpatient Clinic assisted with coordination of patient visits. The Hamber Library and the Biomedical Communications departments at B.C. Children's Hospital provided invaluable assistance.

Special thanks are owed to the following individuals: Gayle Quigley who coordinated the pregnancy and control studies, Beryl Jacobson who developed many of the specialized micronutrient and enzyme assays, Jill Mahy, who coordinated enrollment of the pregnancy volunteers, and Stacey Behl, who provided superb secretarial and library assistance.

Several companies provided generous support in the form of reagents or supplies needed to carry out the extensive laboratory analyses required for this undertaking. These included Kodak Canada, Toronto, ON; Behring Canada, Hoechst Diagnostics, Montreal QU; Varian Canada Inc., Georgetown, ON; Pharmacia, Baie d'Urfe, QU; Becton-Dickinson, Mississauga, ON; Abbott Laboratories, Mississauga, ON.

The support of the B.C. Children's Hospital Telethon Project Grants Program is greatly appreciated.

A special thanks must go to Dr. David Hardwick, Director of Laboratories at B.C. Children's Hospital, for creating an environment which supports and encourages scientific investigation.

Most important, I would like to say thank you to my husband, Bob Lockitch, for many years of love, encouragement, and infinite patience, and to my children, Michael, Keith, and Amanda, for being a daily source of pride and joy.

TABLE OF CONTENTS

PART I. GESTATIONAL CHANGES IN NORMAL PREGNANCY BIOCHEMISTRY

PART III. REFERENCE RANGES FOR NON-PREGNANT ADULT WOMEN

PART IV. STUDY PROTOCOL, MATERIALS, AND METHODS

Handbook
of
Diagnostic Biochemistry
and
Hematology
in Normal Pregnancy

Part I

Gestational Changes in Normal Pregnancy

I.A. Physiological Adaptations in Normal Pregnancy

Healthy normal pregnancy is characterized by major physiological and anatomical adaptations which alter maternal blood biochemistry and hematology laboratory values to varying degrees from those of the non-pregnant adult female. An appreciation of these changes is essential to avoid incorrect interpretation of these changes as pathological, to aid in understanding nutritional requirements, and to optimize pharmacotherapy in pregnancy.

A successful pregnancy outcome requires that the mother provides a protective fetal environment with sufficient but not excessive amniotic fluid, effective maternal-to-fetal transport of nutrients, efficient clearance of toxic metabolic products, and adequate gas exchange. In addition, the mother must acquire stores to support herself and her infant through pregnancy and lactation and protect herself against the hazard of blood loss during labor.

Adaptive mechanisms primarily mediated by maternal and placental hormones include early expansion of plasma volume, alterations in cardiac output and distribution of blood flow, increased glomerular filtration, expansion of erythrocyte mass, and stimulation of hepatic protein synthesis. Interaction of these factors dramatically alters blood biochemistry.

Aspects of these adaptive changes in pregnancy have been discussed in several reviews or books.[1-8] The overall effect of these physiological changes on maternal metabolism will be briefly reviewed here. The impact on specific groups of biochemistry or hematology tests will be discussed in the relevant sections.

HORMONAL EFFECTS IN PREGNANCY

Many of the significant changes in plasma concentration of biochemical and hematological indices result from the hormonal changes of pregnancy.[8,9]

Progesterone is required to sustain early embryonic growth. Following conception, the corpus luteum maintains high levels of plasma progesterone until the developing placenta synthesizes sufficient progesterone to sustain pregnancy. Plasma levels of progesterone increase linearly from the first trimester to the third trimester[8] and at term are 5 to 20 times higher than during the luteal phase of the normal menstrual cycle.[9]

Progesterone opposes the effect of estrogen by decreasing the number of estrogen receptors and enhancing estradiol conversion to estrone. The aldosterone effect on renal tubules is inhibited by progesterone, thus decreasing tubular reabsorption of sodium.[9] Progesterone also inhibits smooth muscle tone in the uterus, gastrointestinal tract, gallbladder, and renal collecting system. This decreases uterine contractility, but also increases the tendency to esophageal reflux, delayed gastric emptying, constipation, gallbladder enlargement, and hydronephrosis.[4] Progesterone decreases the sensitivity of arterial smooth muscle fibers to angiotensin II, decreasing smooth muscle tone in the peripheral vascular system and expanding vascular capacity. Progesterone stimulates hyperventilation, thus promoting respiratory alkalosis with a compensating decrease in plasma bicarbonate.

Estradiol levels increase greatly from the first to the third trimester.[8,9] Growth of the fetus stimulates estrogen synthesis through positive feedback as the fetal adrenal production of dehydroepiandrosterone sulfate increases.[9,10] Estrogen in turn stimulates expansion of blood flow to uterus and mammary glands.[8] Estrogen stimulates hepatic synthesis of proteases, coagulation factors, binding proteins, and plasma renin substrate.[11-15] Very high plasma concentrations of protein-bound hormones and nutrients result from the estrogen effect on the liver.

Although many of the major adaptive changes result from the effects of progesterone and estrogen, alterations in the metabolism of protein, fat, and carbohydrate are also brought about by increases in other steroid hormones and the interaction of insulin with insulin antagonists.[7,8]

The placenta is a major source of both steroid and polypeptide hormones. Important placental peptide hormones include human chorionic gonadotropin (HCG), human placental lactogen (human chorionic somatotropin) (HPL), and chorionic corticotropin, thyrotropin, and follicle-stimulating hormone. Luteinizing hormone-releasing factor, thyrotropin-releasing factor, and nerve growth factor are also produced by the chorionic cells.[6-8]

HCG, synthesized by the syncytiotrophoblast, reaches peak concentration around the 10th week of pregnancy and then decreases gradually to level out at about 20 weeks.[8] This decrease appears to be delayed in pregnancies with Down syndrome.[16]

Placental lactogen is a major mediator of the metabolic changes of pregnancy.[7,8] It is probably the most important insulin antagonist, and stimulates the use of lipid as a caloric source, sparing glucose and amino acids for fetal needs.

HEMODYNAMIC EFFECTS IN PREGNANCY

BLOOD VOLUME

Net retention of water and sodium is a primary requirement in pregnancy. A major component of pregnancy weight gain, is due to expansion of maternal total body water, which increases by about 4 to 6 L, of which 75% enters the maternal extracellular compartment.[17] Intracellular water in the expanding cell mass of the uterus, placenta, and the fetus accounts for 25% of the retained fluid.

An average increase in blood volume of approximately 1500 to 1600 mL occurs in the normal singleton pregnancy. Expansion of the erythrocyte mass occurs later than the expansion in plasma volume and is maximal in the third trimester[18] (see Part II). Approximately 300 to 400 mL of the extra blood volume represents an increase in red cell mass while 1200 to 1300 mL is due to increased plasma volume.[18]

This hemodilution with a net decrease in maternal hemoglobin concentration serves to maintain adequate uteroplacental, skin, and renal blood flow.[18] The lowered hemoglobin concentration and diminished blood viscosity also counteract the increased hemostatic tendency associated with a progressive increase in coagulation factors during pregnancy[12] (see Part II).

PLASMA VOLUME

Plasma volume at term is increased by approximately 45% or 1300 mL over the non-pregnant state.[18,19] Expansion of blood volume is required for adequate blood supply to placenta and fetus. The relatively greater increase in plasma over red cell mass facilitates more efficient circulation to the skin for heat loss and to the kidneys for increased excretory functions.[18]

Plasma volume expansion correlates with infant birth weight.[19] The increase in blood volume is greater in multiple than in singleton pregnancies and in multigravida than in primigravida.[18] Failure to increase plasma volume has been associated with poor pregnancy outcome, hypertension, and intrauterine growth retardation.[20,21]

Expansion of plasma volume results in an apparent decrease in concentration of sodium, albumin, and hemoglobin, despite the fact that there is effectively a net increase in total circulating mass of each analyte.

FACTORS PROMOTING PLASMA VOLUME EXPANSION

The mechanism of plasma volume expansion is not completely understood. It is postulated that volume expansion results from interaction between the hormone-mediated increase in the renin-angiotensin-aldosterone axis and the arteriovenous shunt effect of expanded uteroplacental circulation.[10] It is probable that both estrogens and progesterone increase plasma aldosterone. Estrogenic stimulation of the liver results in increased production of renin substrate (angiotensinogen)[11] which is activated by renin to angiotensin I. Pulmonary angiotensin-converting enzyme then converts angiotensin I to angiotensin II which in turn stimulates aldosterone production. Progesterone has also been shown to increase plasma renin activity and urinary aldosterone excretion.[22] Levels of plasma renin substrate, renin activity, angiotensin II, and aldosterone increase early in pregnancy and remain high.[23] Renal prostaglandin E_2 production is stimulated by the renin-angiotensin system.[23] Prostaglandin E_2 opposes the vasopressor effect of angiotensin II.[24] The interaction between progesterone, prostaglandin E_2, and angiotensin II serves to keep the blood pressure of the healthy gravida slightly lower, despite the expansion in blood volume and increased cardiac output.[25]

It is probable that the early stimulus to increased renin-angiotensin activity is lowered peripheral resistance due to progesterone-induced smooth muscle relaxation, development of uterine venous plexuses, and large placental arteriovenous shunts. The lowered peripheral resistance acts through the renin-angiotensin system to stimulate an increase in plasma volume to fill the vascular space.[10] The lowered vascular resistance is associated with a slight lowering of systolic blood pressure (10 to 15 mm Hg), a greater decrease in diastolic pressure (20 to 25 mm Hg), and a wider pulse pressure, maximal during mid-pregnancy.[25] Renin production is stimulated by the tendency toward lower blood pressure. With increasing growth of the fetal-placental unit, increased dehydroepiandrosterone sulfate production by the fetal adrenal gland leads to increased placental production of 17-β-estradiol with stimulation of renin substrate synthesis.[10]

BLOOD FLOW

Cardiac output increases by 30 to 40% from the non-pregnant state with most of the increase going to the uterus and placenta.[4] Increased flow to the skin, extremities, and mammary glands also occurs.[4] Although the percent of total cardiac output to the kidneys remains at about 18%, actual renal blood flow increases by 50% over the non-pregnant state. Absolute hepatic blood flow is not significantly altered, although there is a reduction in the percentage of total cardiac output flowing through the liver.[9,26]

RENAL FUNCTION

Renal plasma flow and glomerular filtration rate (GFR) increase early in the first trimester and by 20 weeks are 50% higher than the non-pregnant state.[27,28] Renal plasma flow estimated by PAH clearance increases from a non-pregnant 500 mL/min to 700 mL/min while GFR estimated by 24-hour creatinine clearance increases from 100 mL/min to 140 to 150 mL/min.[27] Although a confounding effect of posture on studies of renal function in late pregnancy has been recognized,[4,27-29] it is probable that during the third trimester GFR again decreases.[27] An early decrease in filtration fraction occurs, but during the third trimester lowered plasma oncotic pressure and increased efferent arteriole vasoconstriction result in an increased filtration fraction.[4]

With the alterations in plasma volume and renal function, plasma concentrations of both urea and creatinine decrease. Altered tubular reabsorption of glucose, uric acid, proteins, and amino acids is also reflected in altered plasma concentration of these substances.[3,4,27] Early in pregnancy plasma concentrations of urea, creatinine, and uric acid decrease,[3] although uric acid levels increase later as tubular reabsorption of urate increases.[30] Renal excretion of glucose,[3] amino acids,[32] and protein[27] is increased from the non-pregnant state.

HEPATIC FUNCTION

The most important effect of pregnancy on liver function is the estrogen-stimulated hepatic protein synthesis.[9,11-15] This causes marked changes in plasma concentration of coagulation factors,[12] renin substrate (angiotensinogen),[11] hormone-binding proteins,[13,14] and acute phase reactants such as ceruloplasmin[13,14] and α-1-antitrypsin.[15]

Absolute blood flow through the liver is not changed in pregnancy; therefore, the liver actually receives a smaller percentage of cardiac output.[26] The relative decrease in blood volume passed through the liver in combination with increased volume of distribution may affect clearance of some drugs used during pregnancy[9] and reduce hepatic clearance of toxic substances. Bromsulfthalein excretion may be reduced.[9]

In uncomplicated pregnancy there is little effect on plasma activity of liver enzymes or on bilirubin metabolism[26] (Section I.E.).

RESPIRATORY FUNCTION

Respiration is stimulated during the progestational phase of menstruation resulting in a tendency to mild hyperventilation with a relative decrease in alveolar and arterial pCO_2 and relative respiratory alkalosis.[5] When pregnancy occurs, the stimulatory effect of progesterone on the respiratory center persists. The decline in pCO_2 correlates well with levels of progesterone. By 2 days postpartum, with the decreased levels of progesterone, arterial pCO_2 is within non-pregnant limits. The respiratory alkalosis is partially compensated by a reduction in plasma bicarbonate but there is a slight net increase in blood pH.[5]

POSTPARTUM PHYSIOLOGICAL ADAPTATIONS

Following an average blood loss at vaginal delivery of 500 mL, rapid adaptations occur to prevent hypotension and maintain perfusion.[32] The loss of placental and fetal circulation and contraction of uterine venous plexi reduces vascular bed capacity. Plasma volume increases soon after delivery through mobilization of extracellular interstitial fluid from the lower extremities. This results in a further acute decrease in hematocrit and plasma protein concentrations.

By the third postpartum day, plasma volume has decreased by about a liter as a result of diuresis.[32] It is possible that atrial natriuretic factor plays a role in this diuresis. Atrial natriuretic factor in plasma was shown to increase from the first to the third trimester[33] and a further increase in both the C-terminus and N-terminus concentration occurred within the 48 hours after delivery.[33]

At the time of the 6-week postpartum visit plasma volume is within the non-pregnant range and most of the laboratory indices have returned to non-pregnant levels. Exceptions to this include uric acid and high-density lipoprotein cholesterol which remain elevated for several months after delivery.

THE PREGNANCY STUDY

RECRUITMENT OF PREGNANT SUBJECTS

Healthy low-risk pregnant women were referred to the study by their physicians at the first trimester visit. Alcohol or drug abusers, or individuals with known medical disorders, were not recruited. Written consent was obtained and the initial (12 ± 1 week) blood specimen obtained.

Further specimens were collected at approximately 16 to 18, 24, 28, 32, and 36 weeks. Blood was also obtained at term (38 to 42 weeks) when the women were admitted in labor, 1 day postpartum, and at the 6-week postpartum visit.

Following childbirth and completion of sampling for each subject, the clinical data were reviewed for exclusion criteria. Those patients who had delivered a term infant after an uncomplicated pregnancy were included in the analysis. Data from women with a diagnosis of cholestasis, hypertension, complicated pregnancy, or delivery were excluded. Vitamin or mineral supplement intake was noted for each patient but no attempt was made to perform detailed diet histories.

All blood specimens were collected using the standardized collection protocol described in Part IV.B. For the pregnancy group, we found that the stringent requirement for an early morning overnight-fasting sample had to be modified. Our original intent was to obtain only early morning fasting blood specimens to exclude the effect of diurnal variation. However, it rapidly became apparent that many pregnant women visited their physicians in the afternoon and had laboratory testing performed at various times during the day. Since the data were intended to be of *practical* use to physicians and laboratories, blood specimens were deemed acceptable if an interval of at least 4 hours from the preceding meal had passed.

Those women who participated throughout the course of the study were highly motivated, many being physicians, laboratory technologists, nurses, and other professionals. Two were of Oriental background but Canadian born and the rest were of Caucasian background.

FINAL PREGNANCY SAMPLE

Forty-one healthy well-nourished women considered to be low risk were enrolled in the first trimester of pregnancy. Seven women returned for three or less visits and were excluded from the study. A further four women were excluded for one or more complications such as mild cholestasis (three occurrences), mild pregnancy-induced hypertension (two occurrences), gestational diabetes (one occurrence), preterm delivery with mild neonatal respiratory distress syndrome (one occurrence), and one small for gestational age infant with transient tachypnea. Two women were found to have the thalassemia trait and their data are excluded from the hematology and iron study data set.

Data from 29 pregnant women are included in the final report. The mean (sd) age at delivery was 31 (3) years, the mean duration of pregnancy was 40 (1) weeks, and mean birth weight of the infants 3.48 (0.5) kg. The mean Apgar score was 8 at 1 min and 9 at 5 min; 16 women used no nutritional supplements, 4 took only iron supplements, and 9 used multivitamin preparations with or without iron.

SPECIMEN COLLECTION PROTOCOLS, MATERIALS, AND METHODS

The standardized specimen collection and processing protocols and specific test methods and instrumentation used for the pregnancy and control reference range studies are described in Part IV.

REFERENCES

1. **Hytten, F. E. and Lind, T.,** *Diagnostic Indices in Pregnancy,* Ciba-Geigy, Basel, 1973, 1.
2. **National Research Council, Committee on Nutrition of the Mother and Preschool Child,** Laboratory Indices of Nutritional Status in Pregnancy, National Academy of Sciences, Washington, D.C., 1978, 1.
3. **Lind, T.,** Clinical chemistry of pregnancy, *Adv. Clin. Chem.*, 21, 1, 1980.
4. **Elrad, H. and Gleicher, N.,** Physiological changes in pregnancy, in *Principles of Medical Therapy in Pregnancy,* Gleicher, N., Ed., Plenum Press, New York, 1985, 948.
5. **Metcalfe, J., Stock, M. K., and Barron, D. H.,** Maternal physiology during gestation, in *The Physiology of Reproduction,* Knobil, E. and Neill, J., Eds., Raven Press, New York, 1988, 2145.
6. **Rosso, P.,** *Nutrition and Metabolism in Pregnancy,* Oxford University Press, New York, 1990, 1.
7. **Cowett, R. M.,** *Principles of Perinatal-Neonatal Metabolism,* Cowett, R. M., Ed., Springer-Verlag, New York, 1991.
8. **Kase, N. G., Reyniak, J. V., and Bergh, P. A.,** Endocrinology of Pregnancy, in *Complications of Pregnancy: Medical, Surgical, Gynecologic, Psychosocial and Perinatal,* 4th ed., Cherry, S. H. and Merkatz, I. R., Eds., Williams & Wilkins, Baltimore, 1991, 916.
9. **Van Thiel, D. H. and Gavaler, J. S.,** Pregnancy associated steroids and their effects on the liver, *Sem. Liver. Dis.*, 7, 1, 1987.
10. **Longo, L. D.,** Maternal blood volume and cardiac output during pregnancy: a hypothesis of endocrinologic control, *Am. J. Physiol.*, 245, R720, 1983.
11. **Tapia, H. R., Johnson, C. E., and Strong, C. G.,** Effect of oral contraceptive therapy on the renin-angiotensin system in normotensive and hypertensive women, *Obstet. Gynecol.*, 41, 643, 1973.

12. **Stirling, Y., Woolf, L., North, W. R. S., Seghatchian, M. J., and Meade, T. W.,** Hemostasis in normal pregnancy, *Thromb. Haemostas.*, 52, 176, 1984.
13. **Sobbrio, G. A., Granata, A., Granese, D., D'Arrigo, F., Panacea, A., Nicita, R., Pulle, C., and Trimarchi, F.,** Sex hormone binding globulin, cortisol binding globulin, thyroxine binding globulin, ceruloplasmin: changes in treatment with two oral contraceptives low in oestrogen, *Clin. Exp. Obstet. Gynecol.*, 18, 43, 1991.
14. **Song, S., Chen, J.-K., He, M.-L., and Fotherby, K.,** Effect of some oral contraceptives on serum concentrations of sex hormone binding globulin and ceruloplasmin, *Contraception,* 39, 385, 1989.
15. **Barr, A. E., Cole, D. A., and Clague, A. E.,** The effect of an acute phase response or pregnancy on plasma alpha-1-antitrypsin concentrations in persons with various S and Z phenotypes, *Ann. Clin. Biochem.*, 25, 255, 1988.
16. **Chard, T.,** Biochemistry and endocrinology of the Down's syndrome pregnancy, *Ann. N.Y. Acad. Sci.*, 626, 580, 1991.
17. **Cheek, D. B., Petrucco, O. M., Gillespie, A., Ness, D., and Green, R. C.,** Muscle cell growth and the distribution of water and electrolyte in human pregnancy, *Early Hum. Dev.*, 11, 293, 1985.
18. **Hytten, F.,** Blood volume changes in normal pregnancy, *Clin. Hematol.*, 14, 601, 1985.
19. **Pirani, B. B. K., Campbell, D. M., and MacGillivray, I.,** Plasma volume in normal first pregnancy, *J. Obstet. Gynaecol. Br. Commonw.*, 80, 884, 1973.
20. **Goodlin, R. C., Quaife, M. T., and Dirksen, W. J.,** The significance and diagnosis of maternal hypovolemia, *Sem. Perinatol.*, 5, 228, 1981.
21. **Hays, P. M., Cruikshank, D. P., and Dunn, L. J.,** Plasma volume determination in normal and preeclamptic pregnancy, *Am. J. Obstet. Gynecol.*, 151, 958, 1985.
22. **Sundsfjord, J. A.,** Plasma renin activity and aldosterone excretion during prolonged progesterone administration, *Acta Endocrinol.*, 67, 483, 1971.
23. **Symonds, E. M.,** The renin-angiotensin system in pregnancy, *Obstet. Gynecol. Ann.*, 10, 45, 1981.
24. **Gant, N. F., Worley, R. J., Everett, R. B., and McDonald, P. C.,** Control of vascular responsiveness during human pregnancy, *Kidney Int.*, 18, 253, 1980.
25. **Christianson, R. E.,** Studies on blood pressure during pregnancy, *Am. J. Obstet. Gynecol.*, 125, 509, 1976.
26. **Fagan, E. A.,** Disorders of the liver, biliary system and pancreas, in *Medical Disorders in Obstetric Practice,* De Swiet, M., Ed., Blackwell Scientific, Oxford, 1989, 426.
27. **Davison, J.,** Renal Disease, in *Medical Disorders in Obstetric Practice,* De Swiet, M., Ed., Blackwell Scientific, Oxford, 1989, 306.
28. **Winston, J. and Levitt, M. F.,** Renal function, renal disease and pregnancy, in *Complications of Pregnancy: Medical, Surgical, Gynecologic, Psychosocial and Perinatal,* 4th ed., Cherry, S. H. and Merkatz, I. R., Eds., Williams & Wilkins, Baltimore, 1991, 502.
29. **Davison, J. M., Vallotton, M. B., and Lindheimer, M. D.,** Plasma osmolality and urinary concentration and dilution during and after pregnancy: evidence that lateral recumbency inhibits maximal urinary concentrating ability, *Br. J. Obstet. Gynaecol.*, 88, 472, 1981.
30. **Lind, T., Godfrey, K. A., and Otun, H.,** Changes in serum uric acid concentrations during normal pregnancy, *Br. J. Obstet. Gynaecol.*, 91, 128, 1984.
31. **Hytten, F. E. and Cheyne, G. A.,** The aminoaciduria of pregnancy, *J. Obstet. Gynaecol. Br. Commonw.*, 79, 424, 1972.
32. **Ueland, K.,** Maternal cardiovascular dynamics. VII. Intrapartum blood volume changes, *Am. J. Obstet. Gynecol.*, 126, 671, 1976.
33. **Merkouris, R. W., Miller, F. C., Catanzarite, V., Rigg, L. A., Quirk, J. G., and Vesely, D. L.,** Increase in the plasma levels of the N-terminal and C-terminal portions of the prohormone of atrial natriuretic factor during normal pregnancy, *Am. J. Obstet. Gynecol.*, 162, 859, 1990.

I.A.1. GRAPHIC AND TABULAR PRESENTATION OF DATA

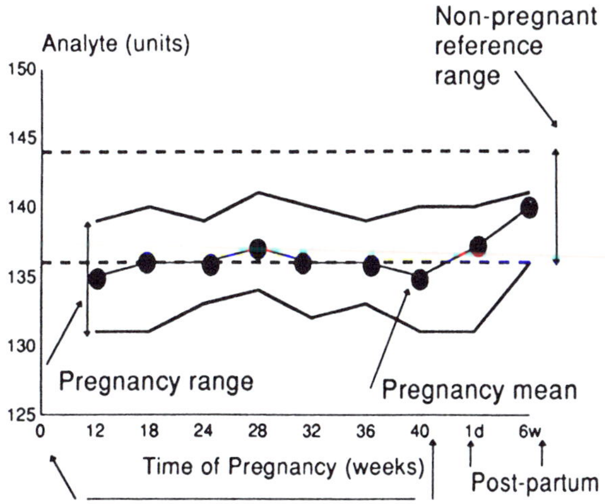

Figure I.A.1. Explanation of graphic data format for each analyte. Note that the X-axis is not to scale.

Table I.A.1. Explanation of tabular data format for each analyte.

Analyte (units)			
Study group (control or time of gestation for pregnancy group)	Reference Range: For controls based on 2.75 and 97.5 fractiles For pregnancy based on mean ± 2 sd	Mean (sd)	Change in pregnancy group mean at each gestational stage compared to non-pregnant control mean
pregnancy (weeks) from last menstrual period			
PP = 1 day or 6 weeks postpartum			

I.B. Fluids, Electrolytes, and Nitrogen Metabolites

EFFECT OF PREGNANCY ON FLUIDS, ELECTROLYTES, AND NITROGEN METABOLITES

The physiological increase in plasma volume and alterations in renal blood flow, filtration, and tubular reabsorption interact to produce marked changes in plasma concentration of many analytes.[1-5] An increase in plasma volume is apparent by 6 to 8 weeks and plasma volume expands progressively reaching a plateau at around 30 to 34 weeks.[6,7] This plasma volume expansion is reflected in lowered concentrations of many substances, including electrolytes such as sodium and proteins such as albumin and prealbumin.[1-5] Associated with the lower albumin concentration, total calcium concentration is decreased though ionized calcium is minimally affected.[8] Lower concentrations of other substances including micronutrients such as zinc[9] may also partially be due to plasma volume expansion.

Renal plasma flow and glomerular filtration rate (GFR) increase by the second trimester to about 50% higher than in the non-pregnant state and decrease slightly in the third trimester.[10,11] The increase in glomerular filtration results in a higher urinary loss of several substances compared with the non-pregnant state. Glucose, amino acids, protein, and several water-soluble vitamins including folic acid are all excreted at greater than usual rates.[4,10,11] Increased clearance of urea, creatinine, and uric acid is also evident early in the first trimester.[3,4,10,11]

Water retention results in an 8 to 10-mosmol/kg decrease in plasma osmolality in early pregnancy.[12] The mean osmolality decreases from 285 $\pm$ 4 to 278 $\pm$ 4 mosmol/kg.[12] Changes in the osmotic threshold for release of arginine vasopressin and for the thirst trigger which occur early in the first trimester[13] alter the normal diuretic response to lowered plasma osmolality.

SODIUM

Approximately 500 to 900 mmol of additional sodium are retained over the course of a normal pregnancy, of which about 60% is lost with the delivery of the infant and placenta.[11] The filtered sodium load may increase by 50% to about 35,000 mmol/day. A net retention of 3 to 4 mmol/day is required to fulfil pregnancy requirements, and thus must be achieved by increased renal tubular sodium reabsorption.[14]

Despite net sodium retention, the effect of expanded plasma volume is shown by a decrease in plasma sodium concentration of 3 to 5 mmol/L in early pregnancy. The sodium concentration remains low relative to the non-pregnant state until after delivery, but by the 6th week postpartum visit normal non-pregnant values are seen.

POTASSIUM

A total retention of about 300 to 350 mmol or about 1 mmol/day of potassium is required over the course of pregnancy.[4] Plasma potassium concentration is tightly regulated within a narrow homeostatic range, and is virtually unchanged throughout pregnancy from the non-pregnant range.

CHLORIDE, BICARBONATE, AND THE ANION GAP

Although bicarbonate levels decrease to compensate for the lowered PCO_2 resulting from increased maternal ventilation, chloride concentrations are unchanged from the non-pregnant state and the anion gap shows a net increase of 1 to 2 mmol/L. The decrease in buffering capacity makes the pregnant patient more susceptible to the effects of ketoacidosis or lactic acidosis.[4] All values have returned to within the non-pregnant reference range by the 6th week postpartum visit.

UREA AND CREATININE

Increased clearance of urea and creatinine early in the first trimester result in significantly lower reference ranges compared to non-pregnant women.[1-5,10,11] It is important to note that concentrations of urea and creatinine which may appear to be within the normal non-pregnant reference range may be high for pregnancy and in fact indicate renal impairment.

The use of plasma creatinine relative to age, height, and weight to estimate GFR cannot be applied in pregnancy where body weight does not reflect kidney size.[10] Both urea and creatinine concentrations are within the non-pregnant reference range by the 6th week postpartum visit.

URIC ACID

Plasma concentrations of uric acid show distinct changes in normal pregnancy resulting from interaction between the changing GFR and tubular reabsorption.[15,16] Uric acid production is not thought to be altered in pregnancy.[16]

From the first trimester until about 28 weeks of gestation a marked decrease occurs in plasma uric acid levels.[15,16] Levels then gradually increase so that plasma uric acid concentrations approximate non-pregnant levels around 32 to 36 weeks. As a result of these increases many women with normal pregnancy have plasma uric acid levels well above the non-pregnant range until term. Both our data and that of Lind[16] showed that plasma uric acid remained elevated above the non-pregnant state at 6 to 12 weeks postpartum.

Recognition of these changes that occur in normal pregnancy is important in evaluating uric acid data in hypertensive and renal disorders in pregnancy.

Thus, a rising uric acid at 20 weeks or a uric acid concentration near the upper limit of the non-pregnant reference range may presage the development of preeclampsia.

DIAGNOSTIC USE OF THESE TESTS IN PREGNANCY

PREECLAMPSIA AND ECLAMPSIA

Preeclampsia has conventionally been recognized by hypertension, edema, and proteinuria occurring most frequently in young primigravida during the second half of pregnancy.[11,17,18] However, preeclampsia is a multisystem disorder of pregnancy which affects the maternal cardiovascular, renal, hepatic, coagulation, and central nervous systems.[17] Thus, patients may present with thrombocytopenia and liver dysfunction prior to developing hypertension and proteinuria.[18] The development of headache, visual disturbances, vomiting, epigastric pain, and neuromuscular irritability implies that full-blown eclampsia with convulsions may rapidly develop. Pulmonary edema associated with severe reduction in plasma colloid osmotic pressure is a dangerous complication of severe renal protein loss.[18]

In the initial phase of preeclampsia the patient may be asymptomatic and identification of this disorder requires screening by blood pressure monitoring and urinalysis. Edema is not a useful predictor. Physiological edema occurs in normal pregnancy[19] and the incidence of hypertension was found to be similar in pregnant women with or without early or late onset edema.[20]

Since neither hypertension nor proteinuria in pregnancy are specific to preeclampsia, other tests may be helpful in both the early identification and prognostic evaluation of the preeclampsia syndrome. Hyperuricemia may occur early in the process[17] while marked proteinuria, coagulopathy with thrombocytopenia, and abnormal liver function are later and ominous signs.[17,18] Marked proteinuria is diagnosed by ≥5 gm protein per 24-hour urine specimen or ≥3+ by semiquantitative dipstick.[18]

Uric acid concentration may be useful in differentiating preeclampsia from chronic hypertension without preeclampsia. In the latter situation, plasma uric acid is not usually elevated.[17] A comparison of women with preeclampsia and chronic hypertension who had similar blood pressures showed that a uric acid of ≥350 μmol/L was associated with a tenfold increase in perinatal mortality in the preeclampsia group.[21] However, as with any laboratory result, use of an absolute value from the literature for a cutoff point may be inappropriate for results obtained from another laboratory. More significant would be the relationship of that value to the appropriate gestational reference limits used by the reporting laboratory. Thus, 350 μmol/L is well within the range of values found at term in our healthy subjects. Sequentially elevated uric acid levels may be of more prognostic value than a single high value.[16]

ACUTE RENAL FAILURE

Acute renal failure is not common in early pregnancy today due to a reduction in complications from septic abortion.[10,11] However, acute renal failure may accompany preeclampsia and abruptio placentae in late pregnancy.[10,11] This syndrome is characterized by a sudden marked decrease in GFR, usually with a decrease in urine output to <400 mL/24 hours, and by rising concentration of plasma urea and creatinine.[10,11]

Assessment of acute renal failure in late pregnancy should include a complete blood count, monitoring of urea, creatinine, electrolytes and osmolality, measurement of plasma glucose, albumin, and amylase, and assessment of liver function, coagulation indices and acid-base status.[10]

CHRONIC RENAL FAILURE

Women with mild to moderate chronic renal failure may undergo successful pregnancy with a good outcome, with a normal infant, and no adverse effect on the kidney disease.[10,11] With a creatinine of ≤124 μmol/L, renal function did not worsen but an increased incidence of small-for-gestational age infants, premature births, and fetal death was noted.[22]

With more severe preexisting renal disease, creatinine 125 to 250 μmol/L, or blood urea of 17.8 mmol/L further deterioration of renal function with pregnancy occurred in the majority of the patients.[23,24] A more recent study noted successful obstetrical outcome in most of the pregnancies in a group of women with creatinine concentrations of 150 to 240 μmol/L but 25% of these women showed decreasing renal function.[25]

Most women with severe renal failure and plasma creatinine higher than 250 μmol/L are anovulatory and pregnancy is unlikely.[10]

REFERENCES

1. **Hytten, F. E. and Lind, T.,** *Diagnostic Indices in Pregnancy,* Ciba-Geigy, Basel, 1973, 1.
2. **National Research Council, Committee on Nutrition of the Mother and Preschool Child,** Laboratory Indices of Nutritional Status in Pregnancy, National Academy of Sciences, Washington, DC, 1978, 1.
3. **Lind, T.,** Clinical chemistry of pregnancy, *Adv. Clin. Chem.,* 21, 1, 1980.
4. **Elrad, H. and Gleicher, N.,** Physiological changes in pregnancy, in *Principles of Medical Therapy in Pregnancy,* Gleicher, N., Ed., Plenum Press, New York, 1985, 948.
5. **Metcalfe, J., Stock, M. K., and Barron, D. H.,** Maternal physiology during gestation, in *The Physiology of Reproduction,* Knobil, E. and Neill, J., Eds., Raven Press, New York, 1988, 2145.
6. **Hytten, F.,** Blood volume changes in normal pregnancy, *Clin. Hematol.,* 14, 601, 1985.
7. **Pirani, B. B. K., Campbell, D. M., and MacGillivray, I.,** Plasma volume in normal first pregnancy, *J. Obstet. Gynaecol. Br. Commonw.,* 80, 884, 1973.

8. **Walters, B. and de Swiet, M.,** Bone disease, disease of the parathyroid glands and some other metabolic disorders, in *Medical Disorders in Obstetric Practice,* de Swiet, M., Ed., Blackwell Scientific, Oxford, 1989, 690.
9. **Hambidge, K. M., Drebs, N. F., Jacobs, M. A., Favier, A., Guyette, L., and Ikle, D. N.,** Zinc nutritional status during pregnancy: a longitudinal study, *Am. J. Clin. Nutr.,* 37, 429, 1983.
10. **Davison, J.,** Renal disease, in *Medical Disorders in Obstetric Practice,* de Swiet, M., Ed., Blackwell Scientific, Oxford, 1989, 306.
11. **Winston, J. and Levitt, M. F.,** Renal function, renal disease and pregnancy, in *Complications of Pregnancy: Medical, Surgical, Gynecologic, Psychosocial and Perinatal,* 4th ed., Cherry, S. H. and Merkatz, I. R., Eds., Williams & Wilkins, Baltimore, 1991, 502.
12. **Davison, J. M., Vallotton, M. B., and Lindheimer, M. D.,** Plasma osmolality and urinary concentration and dilution during and after pregnancy: evidence that lateral recumbency inhibits maximal urinary concentrating ability, *Br. J. Obstet. Gynaecol.,* 88, 472, 1981.
13. **Davison, J. M., Shiells, E. A., Phillips, P. R., and Lindheimer, M. D.,** Serial evaluation of vasopressin release and thirst in human pregnancy, *J. Clin. Invest.,* 81, 798, 1988.
14. **Gallery, E. M. and Brown, M. A.,** Control of sodium excretion in human pregnancy, *Am. J. Kid. Dis.,* IX, 290, 1987.
15. **Dunlop, W. and Davison, J. M.,** The effect of normal pregnancy on the renal handling of uric acid, *Br. J. Obstet. Gynaecol.,* 84, 13, 1977.
16. **Lind, T., Godfrey, K. A., and Otun, H.,** Changes in serum uric acid concentrations during normal pregnancy, *Br. J. Obstet. Gynaecol.,* 91, 128, 1984.
17. **Redman, C. R.,** Hypertension in pregnancy, in *Medical Disorders in Obstetric Practice,* de Swiet, M., Ed., Blackwell Scientific, Oxford, 1989, 249.
18. **Lockwood, C. J.,** Preeclampsia and hypertensive disorders of pregnancy, in *Complications of Pregnancy: Medical, Surgical, Gynecologic, Psychosocial and Perinatal,* 4th ed., Cherry, S. H. and Merkatz, I. R., Eds., Williams & Wilkins, Baltimore, 1991, 476.
19. **Robertson, E. G.,** Oedema in normal pregnancy, *J. Reprod. Fert.,* Suppl. 9, 27, 1969.
20. **Robertson, E. G.,** The natural history of oedema during pregnancy, *J. Obstet. Gynaecol. Br. Commonw.,* 78, 520, 1978.
21. **Redman, C. W. G., Beilin, L. J., Bonnar, J., and Wilkinson, R. H.,** Plasma urate measurements in predicting fetal death in hypertensive pregnancy, *Lancet,* i, 1370, 1976.
22. **Lin, C., Linheimer, M., River, P., and Moawad, A.,** Fetal outcome in hypertensive disorders of pregnancy, *Am. J. Obstet. Gynecol.,* 142, 255, 1982.
23. **Bear, R. A.,** Pregnancy in patients with renal disease; a study of 44 cases, *Obstet. Gynecol.,* 48, 13, 1976.
24. **Kincaid-Smith, K. P., Fairley, K. F., and Bullen, M.,** Kidney disease and pregnancy, *Med. J. Aust.,* 2, 1155, 1967.
25. **Hou, S.,** Pregnancy in women with chronic renal disease, *N. Engl. J. Med.,* 312, 836, 1985.

I.B.1. SODIUM

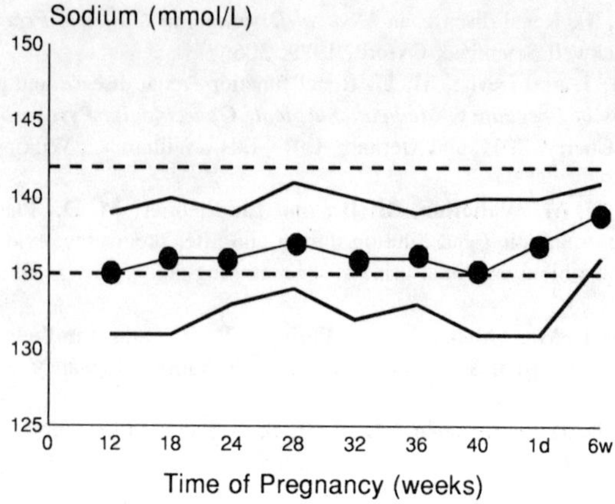

Figure I.B.1. Gestational changes in sodium concentration.

Table I.B.1. Sodium (mmol/L)			
Time of Gestation	Reference Range	Mean (sd)	Percentage of Control mean
Non Pregnant Control	136 - 142	139 (2.0)	
12 weeks	131 - 139	135 (2.0)	97
18 weeks	131 - 140	136 (2.2)	98
24 weeks	133 - 139	136 (1.6)	98
28 weeks	134 - 141	137 (1.7)	99
32 weeks	132 - 140	136 (2.0)	98
36 weeks	133 - 139	136 (1.4)	98
Term	131 - 140	135 (2.2)	97
1 day pp	133 - 141	137 (2.2)	99
6 week pp	138 - 142	140 (1.2)	101

I.B.2. POTASSIUM

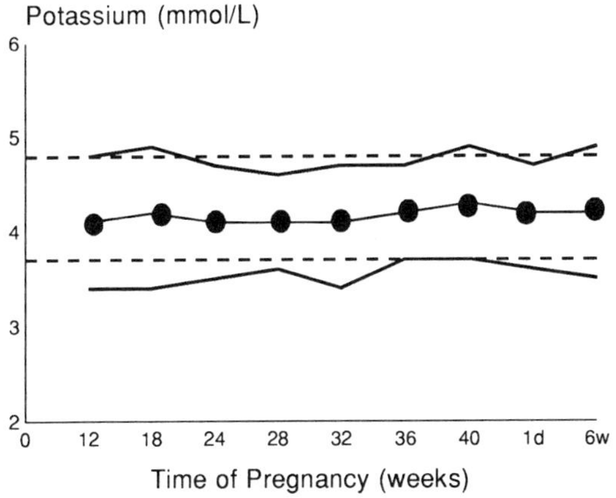

Figure I.B.2. Gestational changes in potassium concentration

Table I.B.2. Potassium (mmol/L)			
Time of Gestation	Reference Range	Mean (sd)	Percentage of Control mean
Non Pregnant Control	3.7 - 4.8	4.3 (0.3)	
12 weeks	3.4 - 4.8	4.1 (0.35)	95
18 weeks	3.4 - 4.9	4.2 (0.38)	98
24 weeks	3.5 - 4.7	4.1 (0.30)	95
28 weeks	3.6 - 4.6	4.1 (0.24)	95
32 weeks	3.4 - 4.7	4.1 (0.33)	95
36 weeks	3.7 - 4.7	4.2 (0.24)	98
Term	3.7 - 4.9	4.3 (0.30)	100
1 day pp	3.6 - 4.7	4.2 (0.28)	98
6 week pp	3.5 - 4.9	4.2 (0.34)	98

I.B.3. CHLORIDE

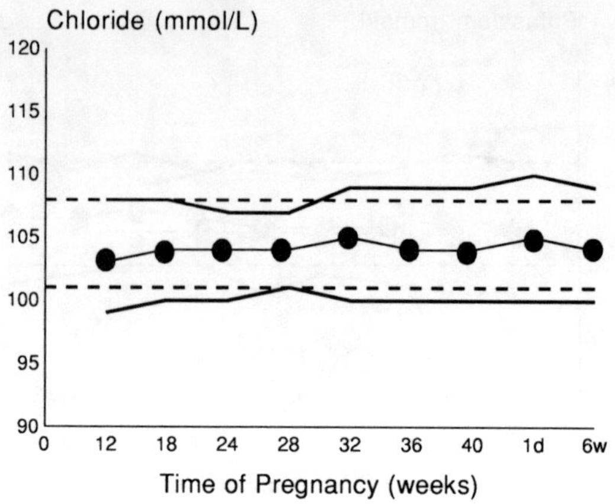

Figure I.B.3. Gestational changes in chloride concentration.

Table I.B.3. Chloride (mmol/L)

Time of Gestation	Reference Range	Mean (sd)	Percentage of Control mean
Non Pregnant Control	101 - 108	105 (1.7)	
12 weeks	99 - 108	103 (2.2)	98
18 weeks	100 - 108	104 (2.1)	99
24 weeks	100 - 107	104 (1.7)	99
28 weeks	101 - 107	104 (1.6)	99
32 weeks	100 - 109	105 (2.1)	100
36 weeks	100 - 108	104 (2.1)	99
Term	100 - 109	104 (2.5)	99
1 day pp	100 - 110	105 (2.5)	100
6 week pp	100 - 108	104 (2.1)	99

I.B.4. BICARBONATE

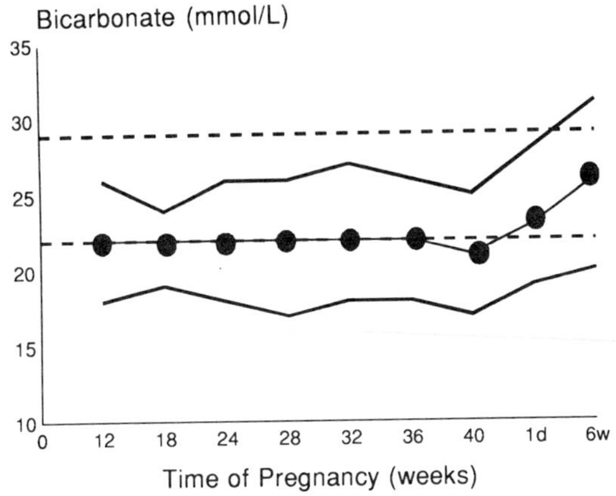

Figure I.B.4. Gestational changes in bicarbonate concentration.

Table I.B.4. Total CO_2 (Bicarbonate) (mmol/L)

Time of Gestation	Reference Range	Mean (sd)	Percentage of Control mean
Non Pregnant Control	22 - 29	26 (1.7)	
12 weeks	18 - 26	22 (2.0)	85
18 weeks	19 - 24	22 (1.3)	85
24 weeks	18 - 26	22 (2.1)	85
28 weeks	17 - 26	22 (1.9)	85
32 weeks	18 - 27	22 (2.2)	85
36 weeks	18 - 26	22 (2.0)	85
Term	17 - 25	21 (2.0)	81
1 day pp	19 - 28	23 (2.3)	88
6 week pp	20 - 31	26 (2.6)	100

I.B.5. ANION GAP

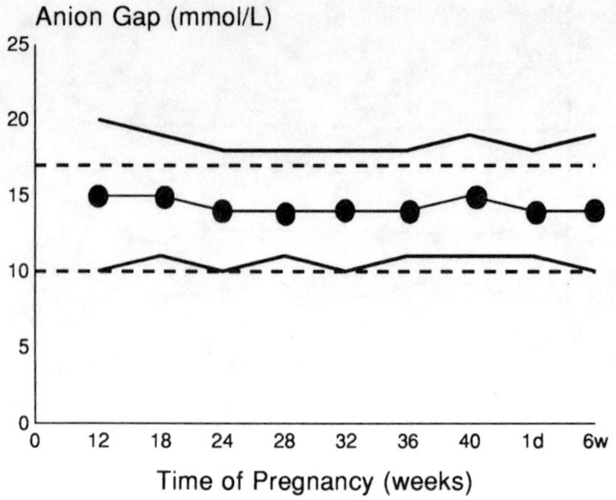

Figure I.B.5. Gestational changes in anion gap concentration.

Table I.B.5. Anion Gap (mmol/L). This is a derived value:
[Anion Gap = (Sodium + Potassium) - (Chloride+Bicarbonate)]

Time of Gestation	Reference Range	Mean (sd)	Percentage of Control mean
Non Pregnant Control	10 - 17	13 (1.6)	
12 weeks	10 - 20	15 (2.4)	115
18 weeks	11 - 19	15 (1.9)	115
24 weeks	10 - 18	14 (2.1)	108
28 weeks	11 - 18	14 (1.8)	108
32 weeks	10 - 18	14 (2.0)	108
36 weeks	11 - 18	14 (1.8)	108
Term	11 - 19	15 (2.1)	115
1 day pp	11 - 18	14 (2.2)	108
6 week pp	10 - 19	14 (2.3)	108

I.B.6. UREA

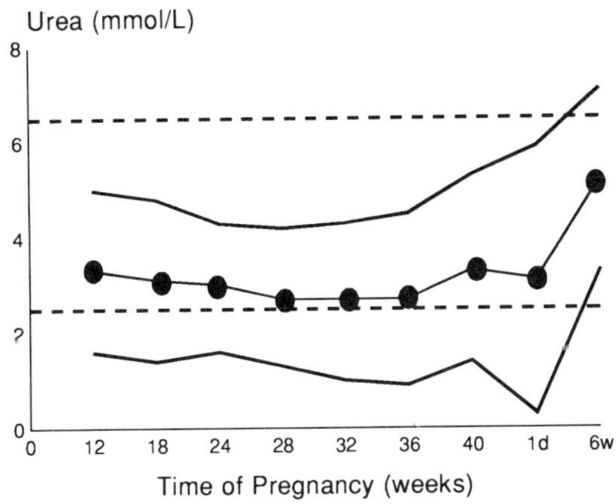

Figure I.B.6. Gestational changes in urea concentration.

Table I.B.6. Urea (mmol/L)			
Time of Gestation	Reference Range	Mean (sd)	Percentage of Control mean
Non Pregnant Control	2.5 - 6.5	4.3 (1.1)	
12 weeks	1.6 - 5.0	3.3 (0.8)	77
18 weeks	1.4 - 4.8	3.1 (0.8)	72
24 weeks	1.6 - 4.3	3.0 (0.7)	70
28 weeks	1.3 - 4.2	2.7 (0.7)	63
32 weeks	1.0 - 4.3	2.7 (0.8)	63
36 weeks	0.9 - 4.5	2.7 (0.9)	63
Term	1.4 - 5.3	3.3 (1.0)	77
1 day pp	0.3 - 5.9	3.1 (1.4)	72
6 week pp	3.3 -7.1	5.2 (1.0)	121

I.B.7. CREATININE

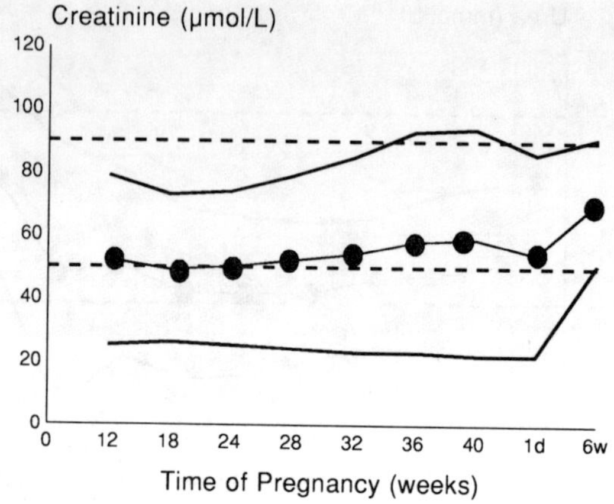

Figure I.B.7. Gestational changes in creatinine concentration.

Table I.B.7. Creatinine (µmol/L)

Time of Gestation	Reference Range	Mean (sd)	Percentage of Control mean
Non Pregnant Control	50 - 90	73 (11)	
12 weeks	25 - 79	52 (13)	71
18 weeks	26 - 73	49 (12)	67
24 weeks	25 - 74	50 (12)	68
28 weeks	24 - 79	52 (14)	71
32 weeks	23 - 85	54 (15)	74
36 weeks	23 - 93	58 (17)	79
Term	22 - 94	59 (18)	81
1 day pp	22 - 86	54 (16)	74
6 week pp	50 - 90	70 (10)	96

I.B.8. URIC ACID

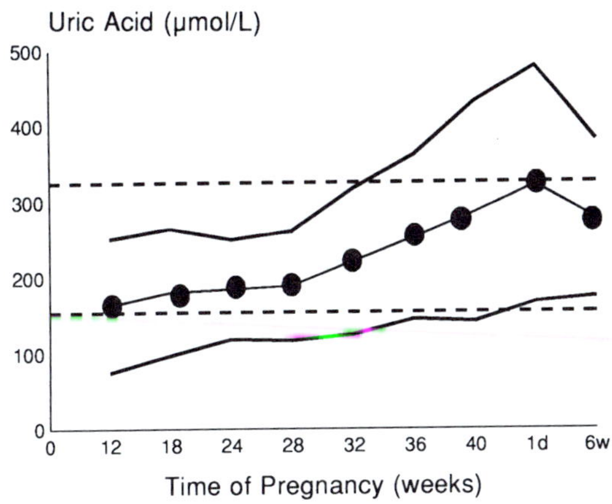

Figure I.B.8. Gestational changes in uric acid concentration.

Table I.B.8. Uric Acid (µmol/L)			
Time of Gestation	Reference Range	Mean (sd)	Percentage of Control mean
Non Pregnant Control	154 - 325	238 (50)	
12 weeks	75 - 251	163 (44)	68
18 weeks	97 - 264	180 (42)	76
24 weeks	118 - 250	184 (33)	77
28 weeks	116 - 261	188 (36)	79
32 weeks	124 - 316	220 (48)	92
36 weeks	144 - 360	252 (54)	106
Term	141 - 429	285 (72)	120
1 day pp	166 - 476	321 (77)	135
6 week pp	173 - 379	276 (52)	116

I.C. Calcium, Phosphorus, and Magnesium

CALCIUM AND PHOSPHORUS METABOLISM

Ninety-nine percent of total body calcium is located in the skeleton, as extracellular hydroxyapatite. The remaining 1%, however, is required for a variety of physiological processes, including coagulation, muscle contraction, nerve conduction, enzyme activation, membrane transport, and hormonal functions. Within the intravascular compartment, about 45% of calcium is protein bound (80% to albumin, 20% to globulins), 10% is bound to other small diffusible ligands such as citrate, bicarbonate, phosphate, and lactate, and about 45% is in the ionized metabolically active form.[1-3]

The concentrations and relative proportions of total and ionized calcium vary with plasma protein concentration and blood pH. Acidosis increases the ionized fraction. The plasma concentration of total calcium decreases with hypoalbuminemia.[1] However, if the ionized fraction remains unchanged, despite the apparent hypocalcemia, no adverse metabolic effect such as tetany will occur.

Serum calcium is homeostatically controlled to maintain the mean total calcium around 2.25 to 2.65 mmol/L and ionized calcium at about half that concentration, about 1.1 to 1.4 mmol/L.[2] This occurs through interaction of parathyroid hormone, calcitonin, and vitamin D affecting intestinal absorption, renal tubular reabsorption, and bone metabolism.[1,2]

Hypocalcemia results from a number of inherited and acquired metabolic disorders due to deficiency of parathyroid hormone or vitamin D.[3] Hypoparathyroidism is due to deficient parathormone secretion and pseudohypoparathyroidism from impaired end organ response to the hormone.[3] Vitamin D deficiency is particularly important in malabsorption disorders, and also occurs in populations that lack sunlight exposure.[4]

Hypercalcemia may result from hyperparathyroidism, excess vitamin D ingestion, thiazide diuretic use, chronic renal failure, malignancy, sarcoidosis, and other hormonal or familial states.[3]

Phosphate is distributed fairly evenly between intracellular and extracellular compartments.[1,2] About 85% of extracellular phosphate occurs as hydroxyapatite in bone. Intracellular phosphate is a component of phospholipid, an important structural component of cellular and subcellular membranes. Intracellular inorganic phosphate is a key component of high-energy transfer reactions.

Phosphate deficiency may result from inadequate oral intake, inadequately supplemented parenteral fluids, renal tubular dysfunction syndromes, and hyperparathyroidism.[1] The most common cause of hyperphosphatemia is acute or chronic renal failure. Metabolic acidosis and vitamin D excess may also elevate serum phosphate.[1] Unlike serum calcium, phosphate concentrations decrease postprandially.[2] Assessment of serum phosphate should preferably be based on a fasting early morning specimen.[2]

EFFECT OF PREGNANCY ON CALCIUM AND PHOSPHORUS METABOLISM

Pregnancy places a greatly increased demand on maternal calcium and phosphate homeostasis.[2,5,6] A term infant contains about 30 g calcium, mostly accumulated in the third trimester.[6] Intrauterine fetal accretion rates for calcium and phosphorus range between 114 to 132 mg/kg/day and 65 to 80 mg/kg/day from 25 to 36 weeks of gestation.[5]

An additional intake of 500 mg/day calcium over normal requirements for a total of about 1300 mg/day and of an additional 200 mg phosphorus for a total of 1250 mg/day were proposed as recent Recommended Nutrient Intakes.[7] The recent Recommended Dietary Allowances suggest a daily intake of 1200 mg of both calcium and phosphorus throughout pregnancy.[8]

Alterations in maternal physiology must occur to optimize maternal-fetal transfer of calcium and phosphate, while maintaining healthy maternal homeostasis. Maternal intestinal calcium retention increases as early as 20 weeks of pregnancy and is almost double that of non-pregnant females by 25 weeks.[9] Maternal total calcium levels decrease throughout pregnancy partially due to the dilutional decrease in serum albumin.[2,10] Ionized calcium concentration has been reported as unchanged, minimally increased, and minimally decreased.[10-12] Fetal calcium levels at term are consistently higher than maternal levels by about 0.25 to 0.5 mmol/L for both total and ionized calcium.[13-15]

Maternal 1,25-dihydroxycholecalciferol increases in pregnancy, partially due to the increase in vitamin D binding protein.[16,17] The changes in maternal parathyroid hormone and calcitonin levels are more controversial.[18,19]

PARATHYROID DISORDERS IN PREGNANCY

Primary hyperparathyroidism is not commonly recognized in pregnancy, but is an important cause of both maternal and fetal morbidity and mortality.[20-22] Even asymptomatic hyperparathyroidism may result in an increased incidence of stillbirth, neonatal death, or spontaneous abortion.[23] By 1988, approximately 100 cases had been reported.[21] It is probable that occult cases are missed since screening for asymptomatic hypercalcemia is not routine practice. Dilutional hypoalbuminemia may mask hypercalcemia; therefore, ionized calcium should be measured.

The disorders associated with hyperparathyroidism may range from a severe life-threatening disorder to asymptomatic maternal hypercalcemia manifested only by neonatal tetany in the infant.[23,24] The risks associated with even occult hyperparathyroidism, and the success of surgical management,[22] make it important to diagnose the condition as early as possible.

The diagnosis should be suspected in women with a history of unexplained miscarriage, threatened abortion, or neonatal tetany. Mild symptoms are non-specific, i.e., nausea and vomiting, tiredness, mild lethargy, and muscle weakness, as well as irritability and depression. Severe cases may present with kidney stones and severe renal colic, pancreatitis, emaciation, thirst, nausea, and skeletal pains.[22]

Hypercalciuria and hypercalcemia, high alkaline phosphatase, and low or normal serum phosphate and magnesium may be found. The total calcium may be normal or only slightly elevated if albumin is very low. Parathormone levels are high. Symptomatic cases may show subperiosteal erosions, bone resorption and cyst formation on radiographs of hand or skull, or nephrolithiasis. Removal of the adenoma in the second trimester is the recommended form of therapy.[22]

Hypoparathyroidism in pregnancy is a rare cause of maternal or fetal problems but can have severe implications for the neonate.[25-27] Pregnancy and lactation in hypoparathyroid women often result in improvement of symptoms such as tetany and in most cases no adverse fetal effects of maternal hypoparathyroidism have been reported. In rare circumstances unrecognized or untreated hypoparathyroidism in the mother may result in fetal calcium levels which are too low to suppress fetal parathormone secretion. This may result in fetal parathyroid hyperplasia characterized by hypercalcemia at birth and skeletal lesions characteristic of hyperparathyroidism.[25-27] Sixteen cases of congenital hyperparathyroidism secondary to maternal hypoparathyroidism were reported by 1990.[27]

EFFECT OF PREGNANCY ON MAGNESIUM METABOLISM

Magnesium is the second-most abundant intracellular cation, with only 1% of total body magnesium in the extracellular fluid compartments. Approximately 60% is contained in the skeleton and almost 40% in muscle and soft tissue.[28] The concentration of magnesium in serum is tightly regulated within the range 0.65 to 1.00 mmol/L.[29] This homeostatic control results from interaction of renal and gastrointestinal mechanisms.[30] Nutritional hypomagnesemia and hypermagnesemia are therefore extremely uncommon in healthy individuals.[31,32]

Approximately one third of serum magnesium is protein bound while two thirds of magnesium is in the ionized, diffusible, and biologically active state.[28] Magnesium is a critically important electrolyte in human physiology, required for activation or modulation of more than 300 enzymes including phosphokinases, phosphatases, thiokinases, and amino acid acyl synthetases.[33-35] Magnesium is required for ribosomal aggregation, RNA binding to ribosomes, and DNA synthesis and degradation.[35] Experimental magnesium deficiency in several animal species resulted in teratogenic effects, abnormalities of parturition, and fetal anemia.[36] Magnesium is required for adenosine 3′,5′-cyclic monophosphate formation.[35] Magnesium deficiency results in impaired sodium-potassium pump efficiency[37] and decreased parathyroid hormone release with resulting hypocalcemia.[37]

Magnesium deficiency is characterized by abnormal neuromuscular signs (positive Trousseau and Chvostek signs, tremors, and muscle spasms), personality disturbance, anorexia, nausea, vomiting, and even coma.[33] Concomitant hypocalcemia and hypokalemia are usually present.[33] Predisposing situations include malabsorptive gastrointestinal disorders, renal tubular dysfunction,[38] diuretic use (furosemide, ethacrynic acid), and unusual inadequate intake such as chronic alcoholism or magnesium-deficient parenteral infusion.[33]

Magnesium toxicity is very uncommon in the absence of renal failure.[33] However, serum magnesium should be carefully monitored in acute and chronic renal insufficiency as levels over 4 mmol/L are potentially lethal.[39]

MAGNESIUM IN PREGNANCY

Normal pregnancy is characterized by a decline in serum magnesium.[36,40,41] Our results show that for most of pregnancy the mean magnesium concentration is at or below the lower limit of the non-pregnant control reference range. An incidence of about 17% of hypomagnesemia (<0.6 mmol/L) has been reported for most studies[40,41] and the existence of subclinical magnesium deficiency in normal pregnancy has been accepted by some authors.[28,40,42] This has been questioned by others.[43,44]

However, from a clinical perspective, symptomatic magnesium deficiency in pregnancy is less well demonstrated. The therapeutic efficacy of magnesium sulfate in management of preeclampsia resulted in speculation that magnesium deficiency may be implicated in this disorder.[40,45] However, women with preeclampsia did not have significantly different concentrations of serum or intracellular magnesium from those without preeclampsia.[45] The use of a 15-mmol daily magnesium supplement given as magnesium aspartate hydrochloride was associated with fewer maternal hospitalizations, reduction in intrapartum hemorrhage, and fewer preterm deliveries.[40] This experience has been duplicated in a recent Hungarian study. Supporting evidence for a positive effect of magnesium supplementation was the reduction in early and later abortion and reduced perinatal mortality in women using magnesium-containing antacids in a large multicenter German study. No toxicity was reported and a reduction in intrauterine growth retardation was noted.[40] A recent study reported a significantly lower serum magnesium in patients in preterm labor without preeclampsia, compared with those not in labor at the same period of gestation.[42] Although serum magnesium was 6% lower in a small group of preeclamptic pregnancies compared with normal pregnancies, the difference was not statistically significant and the authors advised against routine magnesium in pregnant women of good socioeconomic status.[44]

Recent proposals for Recommended Dietary Allowances[31] and Recommended Nutrient Intakes[32] suggested that an increased daily dietary intake of 20 mg or of 15 mg (first trimester) and 45 mg (second and third trimester) were required in pregnancy.

REFERENCES

1. **Fraser, D., Jones, G., Kooh, S. W., and Radde, I. C.,** Calcium and phosphate metabolism, in *Textbook of Clinical Chemistry,* Tietz, N. W., Ed., W. B. Saunders, Philadelphia, 1986, 1317.
2. **Walters, B. and de Swiet, M.,** Bone disease, disease of the parathyroid glands and some other metabolic disorders, in *Medical Disorders in Obstetric Practice,* de Swiet, M., Ed., Blackwell Scientific, Oxford, 1989, 690.
3. **Cole, D. E. C., Carpenter, T. O., and Goltzman, D.,** Calcium homeostasis and disorders of bone and mineral metabolism, in *Pediatric Endocrinology,* 2nd ed., Collu, R., Ducharme, J. R., and Guyda, H. J., Eds., Raven Press, New York, 1989, 509.
4. **Kokkonen, J., Koivisto, M., and Kirkinen, P.,** Seasonal variation in serum-25-OHD3 in mothers and newborn infants in Northern Finland, *Acta Paediatr. Scand.,* 72, 93, 1983.
5. **Ehrenkranz, R. E.,** Mineral needs of the very low birth-weight infant, *Sem. Perinatol.,* 13, 142, 1989.
6. **Pitkin, R. M.,** Calcium metabolism in pregnancy and the perinatal period: a review, *Am. J. Obstet. Gynecol.,* 151, 99, 1985.
7. **Health and Welfare Canada, Scientific Review Committee,** *Nutrition Recommendations,* Canadian Government Publishing Center, Ottawa 1990.
8. **National Research Council, Food and Nutrition Board,** *Recommended Dietary Allowances,* 10th ed., National Academy Press, Washington, DC, 1989.
9. **Heaney, R. P. and Skillman, T. G.,** Calcium metabolism in normal human pregnancy, *J. Clin. Endocrinol.,* 33, 661, 1971.
10. **Payne, R. B., Little, A. J., and Evans, R. T.,** Albumin adjusted calcium concentration in serum increases during normal pregnancy, *Clin. Chem.,* 36, 142, 1990.
11. **Fogh-Anderson, N. and Schultz-Larsen, P.,** Free calcium ion concentration in pregnancy. A longitudinal study on 22 normal pregnant women, *Acta Obstet. Gynecol. Scand.,* 60, 309, 1981.
12. **Gertner, J. M., Coustan, D. R., Kliger, A. S., Mallette, L. E., Ravin, N., and Broadus, A. E.,** Pregnancy as a state of physiologic absorptive hypercalciuria, *Am. J. Med.,* 81, 451, 1986.
13. **Moniz, D. F., Nicolaides, K. H., Bamforth, F. J., and Rodeck, C. H.,** Normal reference ranges for biochemical substances relating to renal, hepatic, and bone function in fetal and maternal plasma throughout pregnancy, *J. Clin. Pathol.,* 38, 468, 1985.
14. **Martinez, M. E., Sanchez, C., Salinas, M., Pellegrini, J., Carracso, A., Catalan, P., Balaguer, G., and Quero, J.,** Ionic calcium levels during pregnancy, at delivery and in the first hours of life, *Scand. J. Clin. Lab. Invest.,* 46, 27, 1986.

15. **Reitz, R. E., Daane, T. A., Woods, J. R., and Weinstein, R. L.,** Calcium, magnesium, phosphorus, and parathyroid hormone interrelationships in pregnancy and newborn infants, *Obstet. Gynecol.*, 50, 701, 1977.
16. **Bouillon, R., Van Assche, F. A., and Van Baelen, H.,** Influence of the vitamin D-binding protein on the serum concentration of 1,25 dihydroxyvitamin D3, *J. Clin. Invest.*, 67, 589, 1981.
17. **Mimouni, F., Tsang, R. C., Hertzberg, V. S., Neumann, V., and Ellis, K.,** Parathyroid hormone and calcitriol changes in normal and insulin-dependent diabetic pregnancies, *Obstet. Gynecol.*, 74, 49, 1989.
18. **Drake, T. S., Kaplan, R. A., and Lewis, T. A.,** The physiologic hyperparathyroidism of pregnancy, *Obstet. Gynecol.*, 53, 746, 1979.
19. **Davis, O. K., Hawkins, D. S., Rubin, L. P., Posillico, J. T., Brown, E. M., and Schiff, I.,** Serum parathyroid hormone (PTH) in pregnant women determined by an immunoradiometric assay for intact PTH, *J. Clin. Endocrinol. Metab.*, 67, 850, 1988.
20. **Shangold, M. M., Dor, N., Welt, S. I., Fleischman, A. R., and Crenshaw, M. C.,** Hyperparathyroidism and pregnancy: a review, *Obstet. Gynecol. Surv.*, 37, 217, 1982.
21. **Higgins, R. V. and Hisley, J. C.,** Primary hyperparathyroidism in pregnancy — a report of two cases, *J. Reprod. Med.*, 33, 726, 1988.
22. **Kelly, T. R.,** Primary hyperparathyroidism during pregnancy, *Surgery,* 110, 1028, 1991.
23. **Better, O. S., Levi, J., Greif, E., Tuma, S., Gellei, B., Erlik, D., and Haifa, I.,** Prolonged neonatal parathyroid suppression, *Arch. Surg.*, 106, 722, 1973.
24. **Jacobsen, B. B., Terslev, E., Lund, B., and Sorensen, O. H.,** Neonatal hypocalcemia associated with maternal hyperparathyroidism, *Arch. Dis. Child.*, 53, 308, 1978.
25. **Aceto, T., Jr., Batt, R. E., Bruck, E., Schultz, R. B., and Perez, Y. R.,** Intrauterine hyperparathyroidism: a complication of untreated maternal hypoparathyroidism, *J. Clin. Endocrinol.*, 26, 487, 1966.
26. **Bronsky, D., Kiamko, R. T., Moncada, R., and Rosenthal, I. M.,** Intra-uterine hyperparathyroidism secondary to maternal hypoparathyroidism, *Pediatrics,* 42, 606, 1968.
27. **Loughead, J. L., Mughal, Z., Mimouni, F., Tsang, R., and Oestreich, C.,** Spectrum and natural history of congenital hyperparathyroidism secondary to maternal hypocalcemia, *Am. J. Perinatol.*, 7, 350, 1990.
28. **Aikawa, J. K.,** *Magnesium: Its Biologic Significance,* Aikawa, J. K., Ed., CRC Press, Boca Raton, FL, 1981, 1.
29. **Lowenstein, F. W. and Stanton, M. F.,** Serum magnesium, *J. Am. Coll. Nutr.*, 5, 399, 1986.
30. **Quamme, G. A.,** Renal handling of magnesium, *Magnesium,* 5, 248, 1986.
31. **National Research Council, Food and Nutrition Board,** *Recommended Dietary Allowances,* 10th ed., National Academy Press, Washington, DC, 1989, 187.
32. **Health and Welfare Canada, Scientific Review Committee,** *Nutrition Recommendations,* Canadian Government Publishing Center, Ottawa, 1990, 136.
33. **Shills, M. E.,** Magnesium in health and disease, *Annu. Rev. Nutr.*, 8, 429, 1988.
34. **Wester, P. O.,** Magnesium, *Am. J. Clin. Nutr.*, 45, 1305, 1987.
35. **Shills, M. E.,** Magnesium, in *Nutrition Review's Present Knowledge in Nutrition,* 5th ed., Brown, M. L., Ed., Nutrition Foundation, Washington, DC, 1984, 422.
36. **Hurley, L. S.,** Magnesium deficiency in pregnancy and its effect on the fetus, *Mag. Bull.*, 1a, 202, 1981.
37. **Berkelhammer, C. and Bear, R. A.,** A clinical approach to common electrolyte problems. IV. Hypomagnesemia, *Can. Med. Assoc. J.*, 132, 360, 1985.
38. **Dudin, K. and Teebi, A. S.,** Primary hypomagnesemia, *Eur. J. Pediatr.*, 146, 303, 1987.
39. **Mordes, J. P. and Wacker, W. A.,** Excess magnesium, *Pharmacol. Rev.*, 29, 273, 1977.

40. **Spatling, L., Disch, G., and Classen, H.-G.,** Magnesium in pregnant women and the newborn, *Mag. Res.*, 2, 4, 1989.
41. **Seelig, M. S.,** Preeclampsia and eclampsia: magnesium levels and treatment, in *Magnesium Deficiency in the Pathogenesis of Disease: Early Roots of Cardiovascular, Skeletal, and Renal Abnormalities,* Seelig, M. S., Ed., Plenum Press, New York, 1980, 38.
42. **Kurzel, R. B.,** Serum magnesium levels in pregnancy and preterm labor, *Am. J. Perinatol.*, 8, 119, 1991.
43. **Hibbard, B.,** Magnesium in pregnancy. Evidence for dietary supplementation is unconvincing, *Br. Med. J.*, 298, 342, 1989.
44. **Skajaa, K., Dorup, I., and Sandstrom, B.-M.,** Magnesium intake and status and pregnancy outcome in a Danish population, *Br. J. Obstet. Gynaecol.*, 98, 919, 1991.
45. **Ryzen, E., Greenspoon, J. S., Diesfield, P., and Rude, R. K.,** Blood mononuclear cell magnesium in normal pregnancy and preeclampsia, *J. Am. Coll. Nutr.*, 6, 121, 1987.

I.C.1. TOTAL CALCIUM

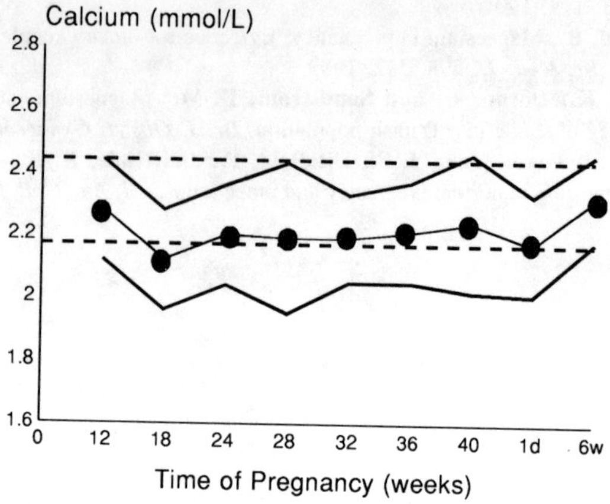

Figure I.C.1. Gestational changes in total calcium concentration

Table I.C.1. Calcium (mmol/L)

Time of Gestation	Reference Range	Mean (sd)	Percentage of Control mean
Non Pregnant Control	2.17 - 2.43	2.32 (0.06)	
12 weeks	2.12 - 2.44	2.28 (0.08)	98
18 weeks	1.96 - 2.28	2.12 (0.08)	91
24 weeks	2.04 - 2.36	2.20 (0.08)	95
28 weeks	1.95 - 2.43	2.19 (0.12)	94
32 weeks	2.05 - 2.33	2.19 (0.07)	94
36 weeks	2.05 - 2.37	2.21 (0.08)	95
Term	2.02 - 2.46	2.24 (0.11)	97
1 day pp	2.01 - 2.33	2.17 (0.08)	94
6 week pp	2.18 - 2.46	2.32 (0.07)	100

I.C.2. IONIZED CALCIUM

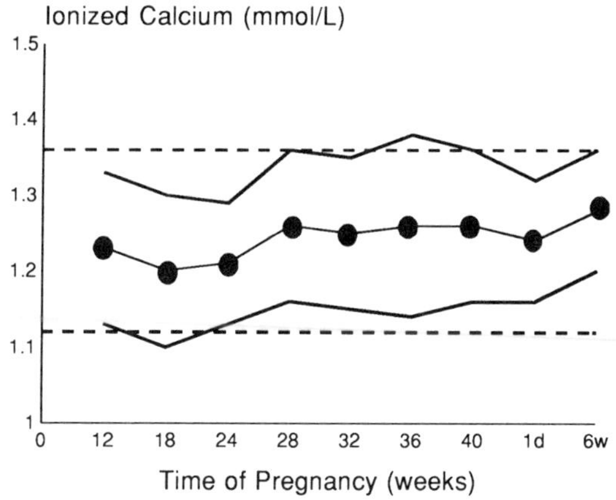

Figure I.C.2. Gestational changes in ionized calcium concentration.

Table I.C.2. Ionized Calcium (mmol/L)

Time of Gestation	Reference Range	Mean (sd)	Percentage of Control mean
Non Pregnant Control	1.10 - 1.36	1.24 (0.05)	
12 weeks	1.13 - 1.33	1.23 (0.05)	99
18 weeks	1.10 - 1.30	1.20 (0.05)	97
24 weeks	1.13 - 1.29	1.21 (0.04)	98
28 weeks	1.16 - 1.36	1.26 (0.05)	102
32 weeks	1.15 - 1.35	1.25 (0.05)	101
36 weeks	1.14 - 1.38	1.26 (0.06)	102
Term	1.16 - 1.36	1.26 (0.05)	102
1 day pp	1.16 - 1.32	1.24 (0.04)	100
6 week pp	1.20 - 1.36	1.28 (0.04)	103

I.C.3. PHOSPHATE

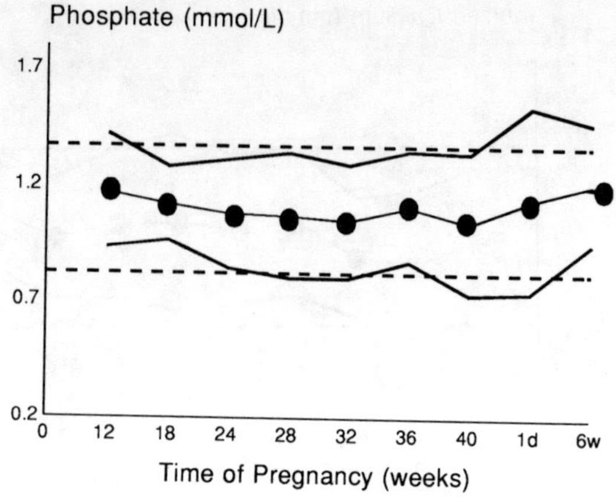

Figure I.C.3. Gestational changes in phosphate concentration

Table I.C.3. Phosphate (mmol/L)			
Time of Gestation	Reference Range	Mean (sd)	Percentage of Control mean
Non Pregnant Control	0.82 - 1.37	1.08 (0.14)	
12 weeks	0.93 - 1.42	1.17 (0.12)	108
18 weeks	0.96 - 1.28	1.12 (0.08)	104
24 weeks	0.84 - 1.31	1.08 (0.12)	100
28 weeks	0.80 - 1.34	1.07 (0.13)	99
32 weeks	0.80 - 1.29	1.05 (0.12)	97
36 weeks	0.87 - 1.35	1.11 (0.12)	103
Term	0.73 - 1.34	1.04 (0.15)	96
1 day pp	0.74 - 1.54	1.14 (0.20)	106
6 week pp	0.95 - 1.47	1.21 (0.13)	112

I.C.4. MAGNESIUM

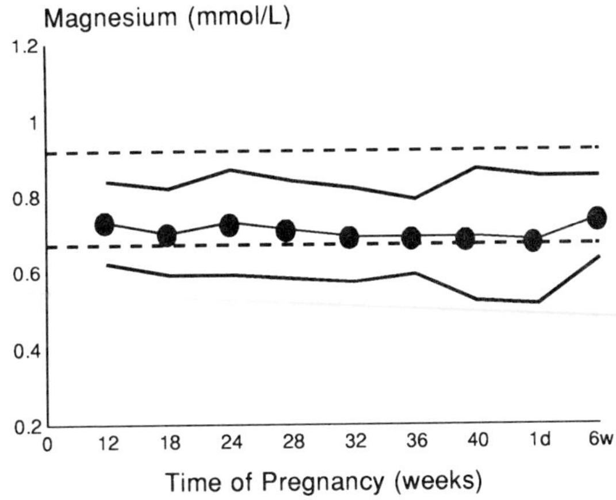

Figure I.C.4. Gestational changes in magnesium concentration

Table I.C.4. Magnesium (mmol/L)

Time of Gestation	Reference Range	Mean (sd)	Percentage of Control mean
Non Pregnant Control	0.67 - 0.92	0.79 (0.06)	
12 weeks	0.62 - 0.84	0.73 (0.06)	92
18 weeks	0.59 - 0.82	0.70 (0.06)	89
24 weeks	0.59 - 0.87	0.73 (0.07)	92
28 weeks	0.58 - 0.84	0.71 (0.07)	90
32 weeks	0.57 - 0.82	0.69 (0.06)	87
36 weeks	0.59 - 0.79	0.69 (0.05)	87
Term	0.52 - 0.87	0.69 (0.09)	87
1 day pp	0.51 - 0.85	0.68 (0.08)	86
6 week pp	0.63 - 0.85	0.74 (0.06)	94

I.D. Glucose

EFFECT OF PREGNANCY ON CARBOHYDRATE METABOLISM

Carbohydrate metabolism is altered in the first weeks of pregnancy by the effects of estrogen, progesterone, and possibly prolactin. Early metabolic changes are anabolic: pancreatic β-cell hyperplasia, increased insulin secretion, and enhanced tissue sensitivity to insulin. Under the influence of insulin, glucose is stored as glycogen and fat, hepatic gluconeogenesis decreases, and utilization and storage of lipids and amino acids are enhanced. Fasting plasma glucose concentrations decrease by 10 to 20%.

The growing placenta contributes significantly to glucose utilization: directly, by metabolizing 50 to 75% of the glucose supplied by the mother and, indirectly, by increasing production of placental lactogen (HPL, hCS) as gestation advances. This hormone, together with cortisol, effects further changes in carbohydrate metabolism in the second half of pregnancy.[1-3]

The most potent diabetogenic hormones, cortisol, progesterone, and HPL, reach peak concentrations in the last half of pregnancy.[1] Hepatic glycogen stores are mobilized, hepatic gluconeogenesis increases, and glucose tolerance in the fed state is stressed. These changes protect the supply of glucose and amino acids to the fetus at the expense of maternal catabolism during fasting. Fasting plasma glucose remains low due to high insulin concentrations, fetal and placental glucose consumption, and urinary glucose losses.[2-4]

About 2 weeks before delivery, glucose utilization increases slightly, for unknown reasons. Energy requirements during labor increase glucose utilization about eightfold, and insulin requirements become negligible.[1] After delivery, parameters of carbohydrate metabolism rapidly change toward expected non-pregnant values,[1,4] particularly in lactating mothers.[5] Insulin requirements and insulin secretion are lower in lactating than in non-lactating mothers.[1]

GLUCOSE

Fasting blood glucose concentrations are approximately 3.3 to 3.9 mmol/L throughout pregnancy;[4] plasma glucose is about 10 to 15% lower.

The blood glucose response to a glucose load depends on the stage of pregnancy. Early in pregnancy the glucose response is blunted compared to the non-pregnant state, but after 20 weeks of gestation 1- and 2-hour postload glucose concentrations increase progressively with gestational age. Weiss has documented reference values for blood glucose response to various glucose loads at different gestational ages.[4] Peak plasma glucose concentration after an oral glucose load is delayed as gestation progresses probably because of delayed gastric emptying.[3,4] Postprandial blood glucose is usually about 7.2 to 7.8 mmol/L.[4]

Lower fasting blood glucose and higher postload glucose concentrations yield mean blood glucose values calculated over 24 hours of about 5.0 to 5.6 mmol/L,[4] or mean plasma glucose concentrations of 4.4 to 5.0 mmol/L[2] in normal pregnancy.

Plasma glucose concentrations in our study were measured at least 4 hours postprandial. Mean concentrations were slightly lower than non-pregnant values at 12 weeks gestation and further decreased 6 to 10% from 18 weeks to term, with a range of 3.4 to 5.5 mmol/L.

Random plasma glucose has been evaluated in England as a screen for gestational diabetes.[6,7] Results in those studies were classified according to whether blood was sampled within 2 hours, or more than 2 hours, after a meal. Our range compares to the reported reference limits for plasma glucose concentrations in samples collected more than 2 hours postprandial: these were 3.5 to 5.4 mmol/L at 10 to 20 weeks gestation, 3.6 to 5.8 mmol/L at 28 to 30 weeks gestation,[6] and <5.6 mmol/L at 28 to 32 weeks gestation.[7] Variability of results increases considerably in samples collected within several hours of a meal.[7] Plasma glucose concentrations up to 6.3 mmol/L are expected in samples collected within 2 hours postprandial.[6]

Random plasma glucose is not currently recommended as a screen for gestational diabetes in North America because it has not yet been directly compared to the 50-g glucose load test advocated here, and reported studies are not comparable.[8,9] "Random" plasma glucose measurements have minimal clinical application; however, glucose levels collected 1 to 2 hours postprandial are useful for monitoring diabetic control.[4,10]

INSULIN

Insulin secretion increases by about 10 weeks of pregnancy. By term, fasting plasma insulin is almost twice the concentration in non-pregnant women, and the insulin response to an intravenous glucose or amino acid load is increased to three to four times the non-pregnant response. Oral glucose or protein loads also invoke an enhanced insulin response in pregnancy, approximately double that in non-pregnant women.[11]

OTHER INDICES OF CARBOHYDRATE METABOLISM OR INSULIN EFFECT

GLYCOSYLATED PROTEINS

Concentrations of glycosylated proteins track the changes in blood glucose concentration during pregnancy, with a lag phase dependent on the protein glycosylated (4 to 8 weeks for hemoglobin A_{1c}, 1 to 3 weeks for fructosamine).[2,8,12]

Hemoglobin A_{1c} concentrations decrease until mid-pregnancy, then increase to reach non-pregnant levels by term.[13,14] Reference ranges are method dependent. Glycosylated hemoglobin is useful for monitoring diabetic control and predicting fetal risk but it lacks sensitivity and specificity as a screen for gestational diabetes.[2,8,9,12] Postpartum measurement of maternal glycosylated hemoglobin has been proposed for retrospective diagnosis of gestational diabetes since glucose tolerance normalizes so rapidly after delivery,[15] but most studies have found it of little use.[16,17]

Glycosylated serum proteins or fructosamine have also been followed during pregnancy. Fructosamine decreases slightly[18,19] or remains constant[20] throughout gestation. As for glycosylated hemoglobin, glycosylated protein measurements seem to be useful for monitoring diabetic control during pregnancy,[12,18,19] but their role in screening for gestational diabetes, despite early enthusiasm,[21] is probably minimal.[17,22,23]

KETONES

Maternal production of ketones and β-hydroxybutyrate increases in late pregnancy as a result of the switch from glucose to fatty acid utilization promoted by the relative lack of insulin (hypoglycemia and hypoinsulinemia during fasting; insulin resistance in the fed state).[24] The placenta also may produce or metabolize ketones or transfer them to the fetus, which uses them as a source of energy, or as substrate for lipid or myelin synthesis.[24,25]

The significance of ketonemia or ketonuria in pregnancy is not clear. In diabetic pregnancies increased serum ketone (β-hydroxybutyrate) concentrations due to inadequate diabetic control have been related to poorer behavioral and intellectual development in the offspring.[26] In contrast, some studies of acetonuria associated with weight loss or low maternal weight gain during pregnancy have failed to show any relation to psychomotor development or IQ up to 7 years of age.[27] Since ketonuria occurs not uncommonly in closely monitored normal pregnancies,[28] its significance clearly requires further study.

LACTATE

Lactate is produced by placental glucose metabolism and is returned to the maternal and fetal circulations. The fetus can use lactate for energy; in the mother it is reconverted to glucose in the liver.[29] Fetal lactate concentrations are higher than maternal. Both maternal and fetal concentrations increase during labor, probably independently. Maternal lactate doubles during labor, due to uterine and skeletal muscle contraction; fetal lactate is related to the severity of fetal hypoxia.[30]

EFFECT OF DISEASE ON BLOOD VALUES

DIABETES

There are three categories of diabetes in pregnancy: insulin-dependent diabetes mellitus (IDDM) recognized before pregnancy, non-insulin-dependent diabetes mellitus (NIDDM) diagnosed before pregnancy, and diabetes diagnosed for the first time during pregnancy (gestational diabetes, GDM). GDM typically resolves after delivery, but some cases represent previously undiagnosed IDDM or NIDDM. The metabolic effect of diabetes on pregnancy varies with the type of diabetes.

Insulin-dependent diabetics lack pancreatic insulin. Exogenous insulin must be increased progressively from the first few weeks of pregnancy to compensate for increased insulin requirements. Strict diabetic control from before conception to the end of the first trimester prevents complications such as congenital anomalies and spontaneous abortion, and possibly macrosomia.[1] Maintenance of fasting plasma glucose concentrations less than 5.5 mmol/L and postprandial plasma glucose less than 6.6 mmol/L throughout pregnancy normalizes concentrations of most metabolic substrates and improves pregnancy outcome.[31]

NIDDM is characterized by increased circulating endogenous insulin and insulin resistance. There is little change in diabetic control in the first trimester of pregnancy because hormonal effects tend to enhance tissue sensitivity to insulin. In the second half of gestation, however, pregnancy-related metabolic changes exacerbate the preexisting insulin resistant state, and requirements for exogenous insulin increase disproportionately. Fasting plasma insulin concentration is higher than in non-diabetic pregnant women, but insulin secretion in response to a stimulus is delayed and daily insulin output is decreased.[1]

GDM is a heterogeneous category. Most women with GDM have sufficient insulin production for non-pregnant requirements, but cannot respond to the additional requirements of pregnancy. Insulin secretion usually resembles the pattern in NIDDM, with increased fasting insulin concentrations, but a decreased response to meals.[1] Most carbohydrate testing in pregnancy is directed toward diagnosing and monitoring GDM, which affects from 1 to 10% of pregnancies.

DIAGNOSIS OF GESTATIONAL DIABETES MELLITUS

Methods for detecting GDM have recently been reviewed.[2,8,9] According to current North American recommendations, all pregnant women should be screened for GDM at 24 to 28 weeks gestation with retesting after 30 weeks in high-risk patients with previously normal results. Abnormal results must be followed by an oral glucose tolerance test for definitive diagnosis.

SCREENING METHODS

Many screening techniques have been proposed, including historical factors, urine glucose, random blood glucose, blood/plasma glucose after an oral glucose, or polysaccharide load, or standard meal, hemoglobin A_{1c} concentration, or plasma fructosamine. Interpretations have been based on venous or capillary, plasma, or whole blood, glucose concentrations determined by a variety of methods. The currently recommended screen is glucose measurement in venous plasma collected 1 hour after a 50-g oral glucose load. Concentrations >7.8 mmol/L are considered abnormal. There is still controversy over the cutoff concentration[2,8,9,32] and whether the patient should be fasting (results may be lower in fed patients).[33] A major problem with the administration of this test is that pregnant patients often vomit the glucose load. Mixed meals or polysaccharide loads may be a better alternative.[8]

DIAGNOSTIC METHODS

As with screening methods, there is controversy over the best test for diagnosing gestational diabetes and the interpretive criteria.[2,8,9,34] The current standard is to measure plasma glucose concentrations fasting and 1, 2, and 3 hours after a 100-g oral glucose load. Patient preparation is important.[9] Two or more values equal to or greater than the following limits are considered abnormal: fasting 5.8 mmol/L; 1 hour 10.6 mmol/L; 2 hour 9.2 mmol/L; 3 hour 8.1 mmol/L.

Lower reference limits have been proposed to correct for changes in glucose assays since this test was first described,[32,34] and other loading doses have been suggested.[4] The importance of two abnormal values has also been questioned, and recent recommendations suggest treating patients with even one abnormally elevated glucose value.[2,8] Development of reference limits for lower doses of glucose or polysaccharide loads may allow evolution of a more acceptable test with fewer side effects such as vomiting.

HYPOGLYCEMIA

Severe hypoglycemia in pregnancy may be a complication of insulin therapy or, very rarely, be due to an insulinoma. Hypoglycemic symptoms of insulinoma may improve[35] or worsen[36] during pregnancy, but typically become more severe postpartum with lactation. Suspected insulinoma in pregnancy is investigated the same way as in a non-pregnant patient, beginning with measurement of insulin:glucose ratios during hypoglycemia or fasting and C-peptide.[36,37]

REFERENCES

1. **Jovanovic-Peterson, L. and Peterson, C. M.,** Glucose metabolism in pregnancy, in *Principles of Perinatal-Neonatal Metabolism,* Cowett, R. M., Ed., Springer-Verlag, New York, 1991, 149.
2. **Dickinson, J. E. and Palmer, S. M.,** Gestational diabetes: pathophysiology and diagnosis, *Sem. Perinatol.,* 14, 2, 1990.
3. **Chahal, P. and Hawkins, D. F.,** Maternal and fetal metabolism: normal and diabetic pregnancies, in *Diabetes and Pregnancy,* Chahal, P. and Hawkins, D. F., Eds., Butterworths, London, 1989, 50.
4. **Weiss, P. A. M.,** Gestational diabetes: a survey and the Graz approach to diagnosis and therapy, in *Gestational Diabetes,* Weiss, P. A. M. and Coustan, D. R., Eds., Springer-Verlag, New York, 1988, 1.
5. **Hubinot, C. J., Balasse, H., Dufrane, S. P., Leclercq-Meyer, V., Sugar, J., Schwers, J., and Malaisse, W. J.,** Changes in pancreatic B cell function during late pregnancy, early lactation and postlactation, *Gynecol. Obstet. Invest.,* 25, 85, 1988.
6. **Hatem, M. and Dennis, K. J.,** A random plasma glucose method for screening for abnormal glucose tolerance in pregnancy, *Br. J. Obstet. Gynaecol.,* 94, 213, 1987.
7. **Lind, T.,** Antenatal screening using random blood glucose values, *Diabetes,* 34, 17, 1985.
8. **Reece, E. A., Assimakopoulos, E., Hagay, Z., Degennaro, N., Zheng, X.-Z., and Hobbins, J. C.,** Assessment of carbohydrate tolerance in pregnancy, *Obstet. Gynecol. Surv.,* 46, 1, 1991.
9. **Cousins, L., Baxi, L., Chez, R., Coustan, D., Gabbe, S., Harris, J., Landon, M., Sacks, D., and Singh, S.,** Screening recommendations for gestational diabetes mellitus, *Am. J. Obstet. Gynecol.,* 165, 493, 1991.
10. **Jovanovic-Peterson, L., Peterson, C. M., Reed, G. F., Metzger, B. E., Mills, J. L., Knopp, R. H., Aarons, J. H., and National Institute of Child Health and Human Development,** Maternal postprandial glucose levels and infant birth weight: the diabetes in early pregnancy study, *Am. J. Obstet. Gynecol.,* 164, 103, 1991.
11. **Kuhl, C.,** Insulin secretion and insulin resistance in pregnancy and GDM, *Diabetes,* 40, 18, 1991.
12. **Peterson, C. M. and Jovanovic, L.,** Glucosylated proteins in normal diabetic pregnancy, in *Gestational Diabetes,* Weiss, P. A. M. and Coustan, D. R., Eds., Springer-Verlag, New York, 1988, 107.
13. **Phelps, R. L., Honig, G. R., Green, D., Metzger, B. E., Frederiksen, M. C., and Freinkel, N.,** Biphasic changes in hemoglobin A_{1c} concentration during normal human pregnancy, *Am. J. Obstet. Gynecol.,* 147, 651, 1983.
14. **Worth, R., Potter, J. M., Drury, J., Fraser, R. B., and Cullen, D. R.,** Glycosylated haemoglobin in normal pregnancy: a longitudinal study with two independent methods, *Diabetology,* 28, 76, 1985.
15. **Pollak, A., Brehm, R., Havelec, L., Lubec, F., Malamitsi-Puchner, A., Simbrunner, G., and Widness, J. A.,** Total glycosylated hemoglobin in mothers of large-for-gestational-age infants: a postpartum test for undetected maternal diabetes?, *Biol. Neonate,* 40, 129, 1981.
16. **Czekelius, P. and Rollmann, J.,** Glycosylated haemoglobins and neonatal macrosomia, in *Gestational Diabetes,* Weiss, P. A. M. and Coustan, D. R., Eds., Springer-Verlag, New York, 1988, 115.
17. **Ghosh, G., Pildes, R. S., Richton, S., and Ajayi, O.,** Maternal and cord serum glycosylated protein in neonatal macrosomia and correlation with birth weight, *Obstet. Gynecol.,* 75, 79, 1990.

18. **Roberts, A. B., Baker, J. R., James, A. G., and Henley, P.,** Fructosamine in the management of gestational diabetes, *Am. J. Obstet. Gynecol.*, 159, 66, 1988.
19. **Frandsen, E. K., Sabagh, T., and Bacchus, R. A.,** Serum fructosamine in diabetic pregnancy, *Clin. Chem.*, 34, 316, 1988.
20. **Staley, M. J. and Murray-Arthur, F.,** Plasma fructosamine in non-diabetic pregnancy, *Br. J. Obstet. Gynaecol.*, 95, 265, 1988.
21. **Roberts, A. B. and Baker, J. R.,** Serum fructosamine: a screening test for diabetes in pregnancy, *Am. J. Obstet. Gynecol.*, 154, 1027, 1986.
22. **Menon, U., Ranjan, M., Jasper, P., and Oommen, A.,** Evaluation of plasma fructosamine as a screening test for gestational diabetes, *Aust. N.Z. J. Obstet. Gynaecol.*, 31, 25, 1991.
23. **Yatscoff, R. W., Mehta, A., and Dean, H.,** Maternal and cord blood glycated albumin and total serum proteins: correlation with maternal glycemic control and birth weight, *Am. J. Obstet. Gynecol.*, 153, 783, 1985.
24. **Rudolf, M. C. J. and Sherwin, R. S.,** Maternal ketosis and its effects on the fetus, *Clin. Endocrinol. Metab.*, 12, 413, 1983.
25. **Hay, W. W., Jr.,** The placenta. Not just a conduit for maternal fuels, *Diabetes*, 40, 44, 1991.
26. **Rizzo, T., Metzger, B. E., Burns, W. J., and Burns, K.,** Correlations between antepartum maternal metabolism and intelligence of offspring, *N. Engl. J. Med.*, 325, 911, 1991.
27. **Naeye, R. L. and Chez, R. A.,** Effects of maternal acetonuria and low pregnancy weight gain on children's psychomotor development, *Am. J. Obstet. Gynecol.*, 139, 189, 1981.
28. **Chez, R. A. and Curcio, F. D., III,** Ketonuria in normal pregnancy, *Obstet. Gynecol.*, 69, 272, 1987.
29. **Munro, H. M., Pilistine, S. J., and Fant, M. E.,** The placenta in nutrition, *Ann. Rev. Nutr.*, 5(3), 197, 1983.
30. **Katz, N., Lunenfeld, E., Mazor, M., and Piura, B.,** Lactate in the human fetus, *J. Perinatal Med.*, 16, 113, 1988.
31. **Reece, E. A., Coustan, D. R., Sherwin, R. S., Tuck, S., Bates, S., O'Connor, T., and Tamborlane, W. V.,** Does intensive glycemic control in diabetic pregnancies result in normalization of other metabolic fuels?, *Am. J. Obstet. Gynecol.*, 165, 126, 1991.
32. **Carpenter, M. W. and Coustan, D. R.,** Criteria for screening tests for gestational diabetes, *Am. J. Obstet. Gynecol.*, 144, 768, 1982.
33. **Coustan, A. D., Widness, J. A., Carpenter, M. W., Rotondo, L., Pratt, D. C., and Oh, W.,** Should the fifty-gram, one-hour plasma glucose screening test for gestational diabetes be administered in the fasting or fed state?, *Am. J. Obstet. Gynecol.*, 154, 1031, 1986.
34. **Sacks, D. A., Abu-Fadil, S., Greenspoon, J. S., and Fotheringham, N.,** Do the current standards for glucose tolerance testing in pregnancy represent a valid conversion of O'Sullivan's original criteria?, *Am. J. Obstet. Gynecol.*, 161, 638, 1989.
35. **Hale, P. J., Hale, J. F., and Nattrass, M.,** Insulinoma and pregnancy. Case report, *Br. J. Obstet. Gynaecol.*, 95, 514, 1988.
36. **Galun, E., Ben-Yehuda, A., Berlatzki, J., Ben-Chetrit, E., and Gross, D. J.,** Insulinoma complicating pregnancy: case report and review of the literature, *Am. J. Obstet. Gynecol.*, 155, 64, 1986.
37. **Rubens, R., Carlier, A., Thiery, M., and Vermeulen, A.,** Pregnancy complicated by insulinoma, *Br. J. Obstet. Gynaecol.*, 84, 543, 1977.

I.D.1. GLUCOSE

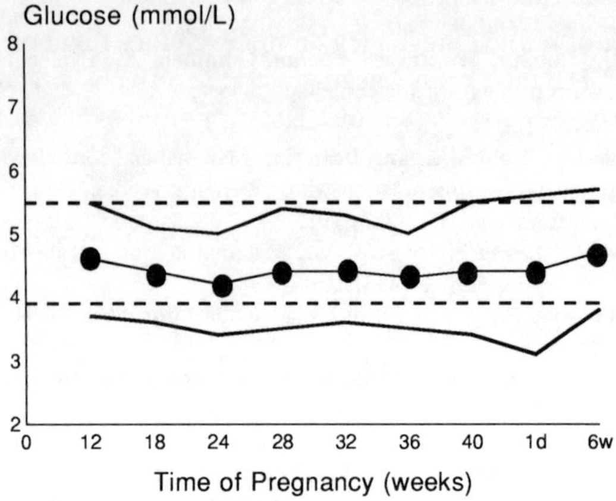

Fig. 1.D.1. Gestational changes in plasma glucose concentration.

Table 1.D. 1. Plasma glucose (mmol/L)			
Time of Gestation	Reference Range	Mean (sd)	% of Control mean
Non Pregnant Control	3.9 - 5.5	4.7 (0.4)	
12 weeks	3.7 - 5.5	4.6 (0.5)	98
18 weeks	3.6 - 5.1	4.4 (0.4)	94
24 weeks	3.3 - 5.0	4.2 (0.4)	89
28 weeks	3.5 - 5.4	4.4 (0.5)	94
32 weeks	3.6 - 5.3	4.4 (0.4)	94
36 weeks	3.5 - 5.0	4.3 (0.4)	91
Term	3.4 - 5.5	4.4 (0.5)	94
1 day pp	3.1 - 5.6	4.4 (0.6)	94
6 week pp	3.7 - 5.7	4.7 (0.5)	100

I.E. Bilirubin and Plasma Enzymes

PHYSIOLOGICAL EFFECT OF PREGNANCY ON HEPATIC FUNCTION AND PLASMA ENZYMES

The effect of normal pregnancy on the excretory functions of the liver is not striking.[1-6] The most pronounced gestational effects on liver biochemistry result from estrogen stimulation of hepatic protein synthesis.[6] Despite a reduction in relative cardiac output to the liver from the non-pregnant state (34% of cardiac output) to about 28%,[6] the actual hepatic blood flow is unchanged and there appears to be minimal reduction in the efficiency of hepatic plasma clearance.[3]

Hepatic cholesterol synthesis and excretion into bile are increased during pregnancy.[7,8] A relatively greater increase in cholesterol synthesis compared to bile acids or phospholipids increases supersaturation of cholesterol in hepatic bile predisposing to cholelithiasis.[6] Increasing hepatic 17-β-glucuronated estrogens are believed to decrease the flow of biliary water, reduce the volume of bile flow, and decrease the secretion of organic anions and solutes.[6] The reduction in hepatic secretory function is shown in the decrease in bromsulfthalein (BSP) clearance although uptake and conjugation are not decreased.[9] A net increase in hepatic storage of BSP occurs.[6]

There are contradictory reports about the effect of normal pregnancy on bilirubin metabolism. Plasma bilirubin concentrations have been reported to be increased, decreased, or unchanged.[4,5,10] Compared with non-pregnant controls, not using oral contraceptives, mean *total bilirubin* was 2 μmol/L lower in women at 36 weeks of pregnancy.[10] We found significantly lower *unconjugated bilirubin* throughout pregnancy compared to the range in our non-pregnant, non-oral contraceptive users. The mean level in pregnancy was 56 to 67% of the control mean. Some of this reduction may be explained by the reduction in plasma albumin concentration, the main transport protein for unconjugated bilirubin. Therefore, although unconjugated bilirubin *concentration* may be low due to the dilutional decrease in albumin, the expanded volume of distribution implies that the total intravascular mass of unconjugated bilirubin is not reduced.

Hepatic protein synthesis is greatly increased for some binding proteins (thyroxine-binding globulin, corticosteroid-binding globulin), transport proteins (ceruloplasmin, transferrin), and coagulation factors (fibrinogen) (see Section I.F.).

Significant changes in plasma enzyme activity during normal pregnancy are not expected for the transaminases (aspartate aminotransferase and alanine aminotransferase), for γ-glutamyl transferase, lactate dehydrogenase, or amylase.[10,11] Our results are in agreement with this. A moderate reduction in mean creatine kinase activity may be noted.[11] It is important to realize that genetic factors may influence plasma enzyme activity[12] and that racial differences have been reported for amylase[13] and creatine kinase.[14]

Alkaline phosphatase increases markedly from around 24 weeks and at term is increased severalfold above the non-pregnant mean.[10,11] The majority of the increase is due to placental production of the heat-stable alkaline phosphatase isoenzyme.[10,11] Alkaline phosphatase may increase with liver or bone disease in pregnancy but isoenzyme studies may be required to differentiate these causes from the natural increase in total alkaline phosphatase activity. Genetic factors may alter plasma alkaline phosphatase activity.[12] Familial hyperphosphatasemia occurs as an autosomal dominant elevation of liver, bone, or intestinal alkaline phosphatase in plasma but is not associated with any known clinical problem.[12]

Serum amylase activity shows little change throughout normal pregnancy.[11] Increased amylase in pregnancy is therefore a useful indicator for pancreatitis, which in pregnancy is often associated with gallstones and cholecystitis.[15,16] However, it has been shown that West-Indian (Afro-Caribbean) and Asian populations have plasma amylase activity that is twice or one and a half times higher than amylase activity in Caucasians.[13] This should be considered to avoid misinterpretation of these elevated levels as suggestive of pancreatitis.

A moderate decrease in total creatine kinase activity occurs early in pregnancy[11] and values remain low until the mid-second or early third trimester. An acute elevation may occur at the time of delivery[14] but levels are usually within the non-pregnant reference range by the 6th week postpartum visit. It has been shown that the increased total creatine kinase is due to increases in both the CK-M fraction, probably from skeletal muscle, and the CK-B fraction, which comes from myometrium, placenta, and fetal blood.[17] Racial differences in plasma total creatine kinase activity have also been reported. Mean plasma creatine kinase activity is about 50% higher in Blacks than in Caucasians or Asians.[14]

Women who are heterozygotes for Duchenne muscular dystrophy may have moderately increased levels of creatine kinase. If creatine kinase is measured in early pregnancy to screen for possible heterozygosity prior to referral for DNA analysis, the usual pregnancy decline in creatine kinase should be considered when interpreting results.

Development of polymyositis or dermatomyositis in pregnancy is not common[18] but women with preexisting myositis may become pregnant.[18] Fetal mortality is high in these cases and, in about one fifth of patients, exacerbation of disease may occur.[18] Creatine kinase levels may be used as an adjunct to monitor changes in severity of polymyositis or dermatomyositis during pregnancy but the propensity of creatinine kinase to decrease in normal pregnancy should again be borne in mind.

With the extreme increase in alkaline phosphatase seen in normal pregnancy, γ-glutamyl transferase may be more useful in the diagnosis of liver disease in pregnancy. The activity of γ-glutamyl transferase varies very little during normal pregnancy.[19] Ethanol ingestion can cause elevated plasma levels of this enzyme.

LIVER DISEASE IN PREGNANCY

It is important to note that clinical signs such as edema, varices, and spider angiomata, which may suggest hepatic disease in the non-pregnant state, may be considered normal when present in the healthy woman in late pregnancy.[20,21] Jaundice is not uncommon in pregnancy with estimates of the incidence ranging from 1:300 to 1:5000 cases.[4,21] Pregnancy is rare in chronic liver disease because of reduced fertility.[21] Approximately 75% of cases of jaundice in pregnancy are due to either acute viral hepatitis or intrahepatic cholestasis of pregnancy.[21] More extensive reviews of liver function and liver disease in pregnancy can be found elsewhere.[1-5,20,21]

HYPEREMESIS GRAVIDARUM

Hyperemesis gravidarum may be associated with major increases in aspartate transaminase,[22] and hyperbilirubinemia.[23] Conversely, lower levels of bilirubin have been reported in emetic pregnancies.[24] Control of vomiting and rehydration rapidly reverses signs of hepatic dysfunction.[22]

PREECLAMPSIA AND ECLAMPSIA

Preeclampsia is a relatively common disorder of pregnancy, particularly but not exclusively affecting primigravida.[25] Where diagnostic criteria are strictly defined, about 9.6% of pregnancies are affected.[26] The combination of rapid weight gain, clinically evident edema, hyperuricemia, proteinuria, and hypertension late in the second trimester, or in the third trimester, indicates preeclampsia.[25,27] Headache, visual disturbance, abdominal pain, heart failure with pulmonary edema, or acute renal failure occur with severe disease.[27] Diagnosis and treatment are critical to prevent the development of seizures and cerebral hemorrhage.[27]

Although central nervous system hemorrhage is the most common cause of death in eclampsia, hepatic complications may account for as much as 10 to 15% of eclampsia-related maternal deaths.[27]

Early recognition requires regular screening by monitoring of blood pressure and urinalysis. Severe proteinuria and hyperuricemia are ominous signs.[25] When preeclampsia is suspected, it is important to evaluate liver function by measuring transaminase and exclude additional end organ involvement (see HELLP syndrome).

HELLP SYNDROME (HEMOLYTIC ANEMIA, ELEVATED LIVER TRANSAMINASES, LOW PLATELETS)

This syndrome may range from a life-threatening situation of severe hypertension in association with hemolytic anemia, elevated liver transaminases, and thrombocytopenia, to a milder degree of hypertension, epigastric pain, raised transaminases, and fetal intrauterine growth retardation.[28,29]

Serial laboratory investigations, including tests of liver function, platelet count, and coagulation parameters, assessed sometimes as frequently as every 4 hours are necessary to keep pace with the sometimes rapid deterioration of the patient's condition. As a rule, within 24 hours of delivery, the test results stabilize or show improvement.

CHOLESTASIS OF PREGNANCY

Some women, with an idiopathic or a hereditary tendency, show in response to the estrogen stimulation of oral contraception or pregnancy a decrease in hepatic excretory function beyond the degree normally seen and become symptomatic.[1,30,31] Generalized pruritis may be the only prominent symptom.

Laboratory evaluation may indicate mild conjugated hyperbilirubinemia, moderate increase in the hepatic isoenzyme of alkaline phosphatase, and elevation of aspartate aminotransferase and alanine aminotransferase. Prothrombin time may be slightly prolonged. Bile acids (cholic acid and chenodeoxycholic acid) are increased 10 to 100 times over levels found in healthy non-pregnant controls, and the increase correlates with the severity of pruritis.[2,30] However, this test is not commonly available to monitor the disease process. In our experience, serial monitoring of the transaminases in symptomatic women has been helpful in identifying the sudden increase in enzyme levels, which is associated with severe and often unbearable maternal pruritis and fetal deterioration. Cholestasis of pregnancy is associated with a high incidence of meconium aspiration[2] and with fetal morbidity and mortality as well as with postpartum hemorrhage in the mother. It is mandatory that the presence of this condition is excluded in all pregnant women with generalized pruritis without rash.

Other conditions characterized by conjugated hyperbilirubinemia may also manifest in pregnancy. Mitochondrial antibodies are elevated in primary biliary cirrhosis, which may present very similarly.[4] Women suspected of intrahepatic cholestasis of pregnancy should be screened for markers of viral hepatitis, which may present in the cholestatic phase.[2] Hepatic ultrasound may be useful to exclude cholelithiasis.[2]

HEPATITIS IN PREGNANCY

Viral hepatitis, due to a wide variety of agents, is the most common cause of jaundice in pregnancy.[2,32] The true frequency of these infections in pregnancy is probably underestimated because anicteric subclinical infection is probably common.[32]

Many of the early symptoms (fatigue, malaise, anorexia, nausea, and vomiting) are common in normal early pregnancy. Darkening of the urine is followed by jaundice and sometimes mild pruritis. An enlarged, tender liver can be palpated.[32]

Alanine and aspartate aminotransferases are usually elevated 10 to 100 times above the non-pregnant state, but hyperbilirubinemia is not usually severe. Serum alkaline phosphatase may be elevated but must be interpreted against the normal increase which occurs in pregnancy.

Transaminases as well as bilirubin and prothrombin time can be used to monitor the course of the disease. Rarely, fulminant hepatic failure may occur. Massive increase in serum bilirubin and the aminotransferases may be accompanied by leukocytosis, hyponatremia, hypokalemia, and hypoglycemia.[32]

In a well-nourished stable population hepatitis is usually well tolerated in pregnancy.[33] However, with the increased incidence of hepatitis in low socioeconomic groups, immigrants, and immune-deficient patients, morbidity and mortality will rise, and close monitoring of liver functions as well as the general health status of mother and fetus are necessary to obtain optimal results.

REFERENCES

1. **Latham, P. S.,** Liver diseases of pregnancy, in *Principles of Medical Therapy in Pregnancy,* Gleicher, N., Ed., Plenum Press, New York, 1985, 832.
2. **Fagan, A. E.,** Disorders of the liver, biliary system and pancreas, in *Medical Disorders in Obstetric Practice,* de Swiet, M., Ed., Blackwell Scientific, Oxford, 1989, 426.
3. **Klion, F. M.,** Liver in normal pregnancy, in *Complications of Pregnancy: Medical, Surgical, Gynecologic, Psychosocial and Perinatal,* 4th ed., Cherry, S. H. and Merkatz, I. R., Eds., Williams & Wilkins, Baltimore, 1991, 797.
4. **Krejs, G. J.,** Jaundice during pregnancy, *Sem. Liver Dis.,* 3, 73, 1983.
5. **Yip, D. M. and Baker, A. L.,** Liver diseases in pregnancy, *Clin. Perinatol.,* 12, 683, 1985.
6. **Van Thiel, D. H. and Gavaler, J. S.,** Pregnancy associated steroids and their effects on the liver, *Sem. Liver Dis.,* 7, 1, 1987.
7. **Feingold, K. R., Wiley, T., Moser, A. H., Lear, S. R., and Wiley, M. H.,** *De novo* cholesterologenesis in pregnancy, *J. Lab. Clin. Med.,* 101, 256, 1983.

8. **Kern, F., Jr., Everson, G. T., DeMark, B., McKinley, C., Showalter, R., Erfling, W., Braverman, D. Z., Szczepanik-van-Leeuwn, P., and Klein, P. D.,** Biliary lipids, bile acids, and gallbladder function in the human female, *J. Clin. Invest.*, 68, 1229, 1981.
9. **Combes, B., Shibata, A., Adams, R., Mitchell, B. D., and Trammell, V.,** Alterations in sulfobromophthalein sodium-removal mechanisms from blood during normal pregnancy, *J. Clin. Invest.*, 42, 1431, 1963.
10. **Knopp, R. H., Bergelin, R. O., Wahl, P. W., Walden, C. E., and Chapman, M. B.,** Clinical chemistry alterations in pregnancy and oral contraceptive use, *Obstet. Gynecol.*, 66, 682, 1985.
11. **Lind, T.,** Clinical chemistry of pregnancy, *Adv. Clin. Chem.*, 21, 1, 1980.
12. **Rosalki, S. B.,** Genetic influences on diagnostic enzymes in plasma, *Enzyme*, 39, 95, 1988.
13. **Tsianos, E. B., Jalali, M. T., Gowenlock, A. H., and Braganza, J. M.,** Ethnic 'hyperamylasemia'. Clarification by isoamylase analysis, *Clin. Chim. Acta*, 124, 13, 1982.
14. **Wong, E. T., Cobb, C., Umehara, M. K., Wolff, G. A., Haywood, L. J., Greenberg, T., and Shaw, S. T.,** Heterogeneity of serum creatine kinase activity among racial and gender groups of the population, *Am. J. Clin. Pathol.*, 79, 582, 1983.
15. **Young, K. R.,** Acute pancreatitis in pregnancy, *Obstet. Gynecol.*, 60, 653, 1982.
16. **McKay, A. J., O'Neill, J., and Imbrie, C. W.,** Pancreatitis, pregnancy and gallstones, *Br. J. Obstet. Gynaecol.*, 87, 47, 1980.
17. **Liras, G., Diaz, V., Alvarez, C., Arenas, J., Sanz, R., and Martinez, V.,** Total creatine kinase (CK) and CK-B activity in maternal blood and cord-blood samples after vaginal and caesarean births, *Clin. Chem.*, 34, 1498, 1988.
18. **Guherrez, G., Dagnino, R., and Mintz, G.,** Polymyositis/dermatomyositis and pregnancy, *Arthr. Rheum.*, 27, 291, 1984.
19. **Walker, F. B., Hoblit, D. L., Cunningham, F. G., and Combes, B.,** Gamma glutamyl transpeptidase in normal pregnancy, *Obstet. Gynecol.*, 43, 745, 1974.
20. **Varma, R. R.,** Course and prognosis of pregnancy in women with liver disease, *Sem. Liver Dis.*, 7, 59, 1987.
21. **Galbraith, R.,** Liver disease: general considerations, in *Principles of Medical Therapy in Pregnancy,* Gleicher, N., Ed., Plenum Press, New York, 1985, 829.
22. **Larrey, D., Rueff, B., Feldman, G., Degott, C., Danan, G., and Benhamou, J. P.,** Recurrent jaundice caused by recurrent hyperemesis gravidarum, *GUT,* 25, 1414, 1984.
23. **Adams, R. H. and Gordon, J.,** Hyperemesis gravidarum. I. Evidence of hepatic dysfunction, *Obstet. Gynecol.*, 31, 659, 1968.
24. **Jarnfeldt-Samsioe, A., Eriksson, B., Waldenstrom, J., and Samsioe, G.,** Serum bile acids, gamma-glutamyltransferase and routine liver function tests in emetic and nonemetic pregnancies, *Gynec. Obstet. Invest.*, 21, 169, 1986.
25. **Redman, C. R.,** Hypertension in pregnancy, in *Medical Disorders in Obstetric Practice,* de Swiet, M., Ed., Blackwell Scientific, Oxford, 1989, 249.
26. **Arngrimsson, R., Steingrimur, B., Geirsson, R. T., Bjornsson, H., Walker, J., and Snaedal, G.,** Genetic and familial predisposition to eclampsia and pre-eclampsia in a defined population, *Br. J. Obstet. Gynecol.*, 97, 762, 1990.
27. **Alexander, J., Cuellar, R. E., and Van Thiel, D. H.,** Toxaemia of pregnancy and the liver, *Sem. Liver Dis.*, 7, 55, 1987.
28. **Weinstein, L.,** Syndrome of hemolysis, elevated liver enzymes, and low platelet count; a consequence of hypertension in pregnancy, *Am. J. Obstet. Gynecol.*, 142, 159, 1982.
29. **Aarnoudse, J. G., Houthoff, H. J., Weits, J., Huisies, V., and Huisies, H. J.,** A syndrome of liver damage and intravascular coagulation in the last trimester of normotensive pregnancy, *Br. J. Obstet. Gynaecol.*, 93, 145, 1986.

30. **Berg, B., Helm, G., Petersohn, L., and Tryding, N.,** Cholestasis of pregnancy, *Acta Obstet. Gynecol. Scand.*, 65, 107, 1986.
31. **Kreek, M. J.,** Female sex steroids and their effects on the liver, *Sem. Liver Dis.*, 7, 8, 1987.
32. **Spechler, S. J. and Koff, R. S.,** Vital hepatitis, in *Principles of Medical Therapy in Pregnancy,* Gleicher, N., Ed., Plenum Press, New York, 1985, 842.
33. **Shalev, E. and Bassan, H. M.,** Viral hepatitis during pregnancy in Israel, *Int. J. Obstet. Gynecol.*, 20, 73, 1982.

I.E.1. BILIRUBIN (UNCONJUGATED)

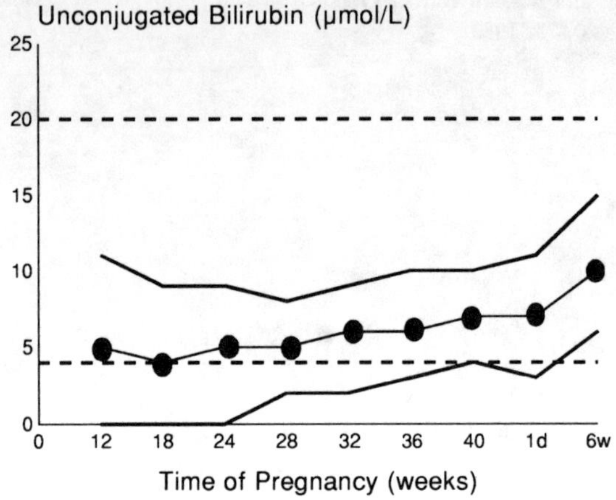

Figure I.E.1. Gestational changes in unconjugated bilirubin concentration.

Table I.E.1. Unconjugated Bilirubin (µmmol/L)			
Time of Gestation	Reference Range	Mean (sd)	Percentage of Control mean
Non Pregnant Control	4 - 20	9 (3.9)	
12 weeks	0 - 11	5 (2.9)	56
18 weeks	0 - 9	5 (2.2)	56
24 weeks	0 - 9	5 (2.1)	56
28 weeks	2 - 8	5 (1.6)	56
32 weeks	2 - 9	6 (1.6)	67
36 weeks	3 - 10	6 (1.7)	67
Term	4 - 10	7 (1.4)	78
1 day pp	3 - 11	7 (2.2)	78
6 week pp	6 - 14	10 (2.1)	111

I.E.2. ALANINE AMINOTRANSFERASE

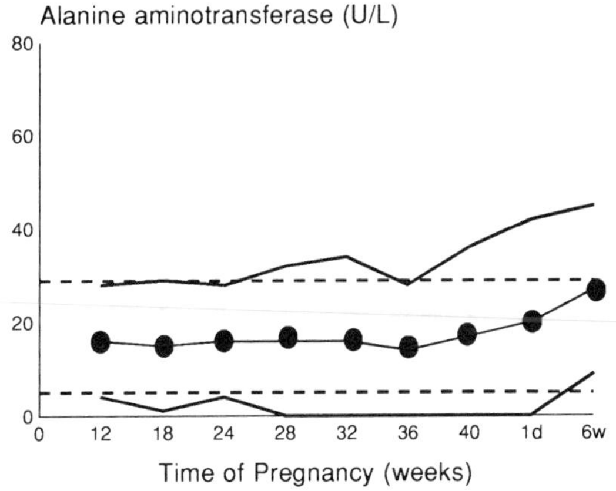

Figure I.E.2. Gestational changes in alanine aminotransferase activity.

Table I.E.2. Alanine aminotransferase (U/L)			
Time of Gestation	Reference Range	Mean (sd)	Percentage of Control Mean
Non Pregnant Control	5 - 29	18 (7)	
12 weeks	4 - 28	16 (6)	89
18 weeks	1 - 29	15 (7)	83
24 weeks	4 - 28	16 (6)	89
28 weeks	0 - 32	16 (8)	89
32 weeks	0 - 34	16 (9)	89
36 weeks	0 - 28	14 (7)	78
Term	0 - 36	17 (9)	94
1 day pp	0 - 42	20 (11)	111
6 week pp	9 - 45	27 (9)	150

I.E.3. ALKALINE PHOSPHATASE

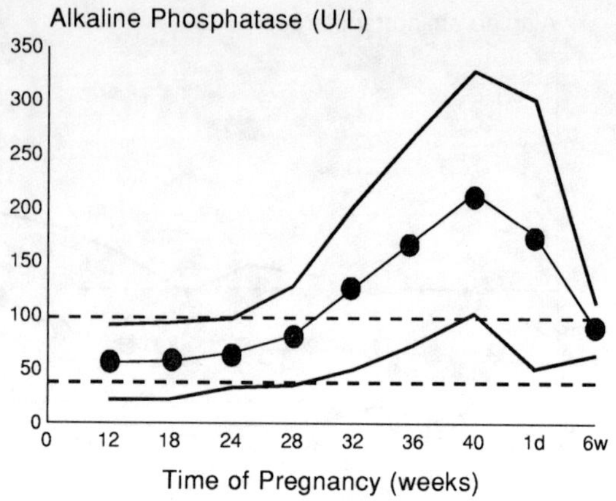

Figure I.E.3. Gestational changes in alkaline phosphatase activity.

Table I.E.3. Alkaline phosphatase (U/L)			
Time of Gestation	Reference Range	Mean (sd)	Percentage of Control mean
Non Pregnant Control	38 - 98	62 (15)	
12 weeks	22 - 91	56 (17)	90
18 weeks	22 - 93	57 (18)	92
24 weeks	33 - 97	65 (16)	105
28 weeks	35 - 127	81 (23)	131
32 weeks	50 - 202	126 (38)	203
36 weeks	73 - 267	170 (48)	274
Term	102 - 328	215 (57)	347
1 day pp	51 - 301	176 (63)	284
6 week pp	64 - 112	88 (12)	142

I.E.4. AMYLASE

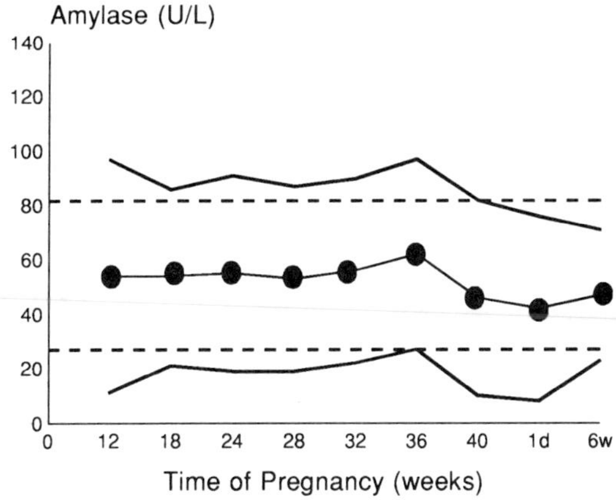

Figure I.E.4. Gestational changes in amylase activity.

Table I.E.4. Amylase (U/L)			
Time of Gestation	Reference Range	Mean (sd)	Percentage of Control mean
Non Pregnant Control	27 - 82	50 (14)	
12 weeks	11 - 97	54 (21)	108
18 weeks	21 - 86	54 (16)	108
24 weeks	19 - 92	55 (18)	110
28 weeks	19 - 87	53 (17)	106
32 weeks	22 - 90	56 (17)	112
36 weeks	27 - 97	62 (17)	124
Term	10 - 82	46 (18)	92
1 day pp	8 - 76	42 (17)	84
6 week pp	23 - 71	47 (12)	94

I.E.5. ASPARTATE AMINOTRANSFERASE

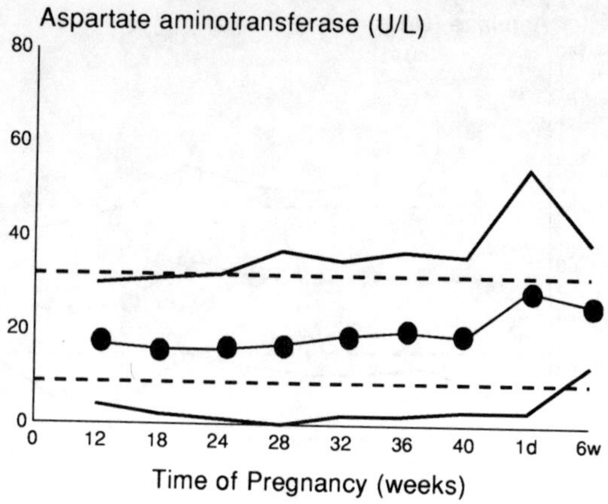

Figure I.E.5. Gestational changes in aspartate aminotransferase activity.

Table I.E.5. Aspartate aminotransferase (U/L)			
Time of Gestation	Reference Range	Mean (sd)	Percentage of Control mean
Non Pregnant Control	9 - 32	20 (6)	
12 weeks	4 - 30	17 (6)	85
18 weeks	2 - 31	16 (7)	80
24 weeks	1 - 32	16 (8)	80
28 weeks	0 - 37	17 (10)	85
32 weeks	2 - 35	19 (8)	95
36 weeks	2 - 37	20 (9)	100
Term	3 - 36	19 (8)	95
1 day pp	3 - 55	29 (13)	145
6 week pp	13 - 39	26 (6)	130

I.E.6. CREATINE KINASE

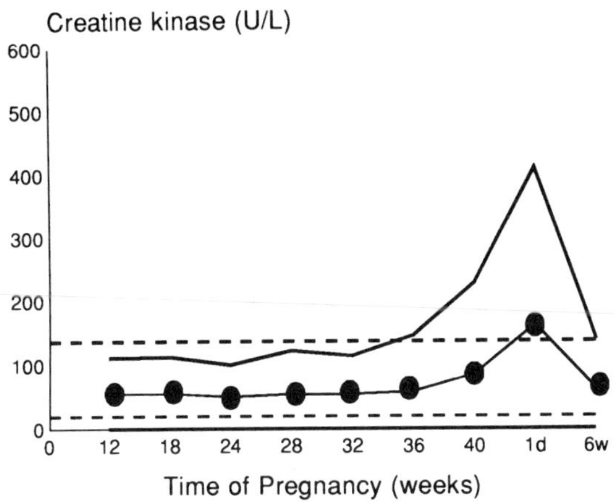

Figure I.E.6. Gestational changes in creatine kinase activity.

Table I.E.6. Creatine kinase (U/L)			
Time of Gestation	Reference Range	Mean (sd)	Percentage of Control mean
Non Pregnant Control	20 - 137	63 (31)	
12 weeks	0 - 111	55 (28)	87
18 weeks	0 - 113	55 (29)	87
24 weeks	0 - 100	50 (25)	79
28 weeks	0 - 122	54 (34)	86
32 weeks	0 - 114	54 (30)	86
36 weeks	0 - 145	57 (44)	90
Term	0 - 227	85 (71)	135
1 day pp	0 - 410	162 (124)	257
6 week pp	0 - 139	63 (37)	100

I.E.7. GAMMA-GLUTAMYL TRANSFERASE

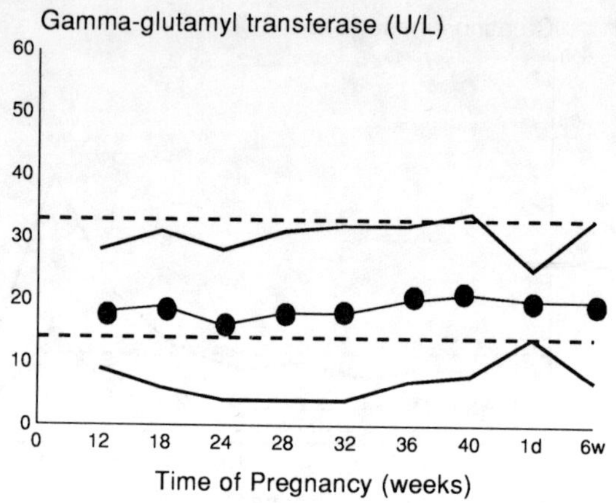

Figure I.E.7. Gestational changes in gamma-glutamyl transferase activity.

Table I.E.7. Gamma-glutamyl transferase (U/L)			
Time of Gestation	Reference Range	Mean (sd)	Percentage of Control mean
Non Pregnant Control	14 - 33	21 (5)	
12 weeks	9 - 28	18 (5)	86
18 weeks	6 - 31	19 (6)	90
24 weeks	4 - 28	16 (6)	76
28 weeks	4 - 31	18 (7)	86
32 weeks	4 - 32	18 (7)	86
36 weeks	7 - 32	20 (6)	95
Term	9 - 34	21 (6)	100
1 day pp	14 - 26	20 (3)	95
6 week pp	7 - 33	20 (7)	95

I.E.8. LACTATE DEHYDROGENASE

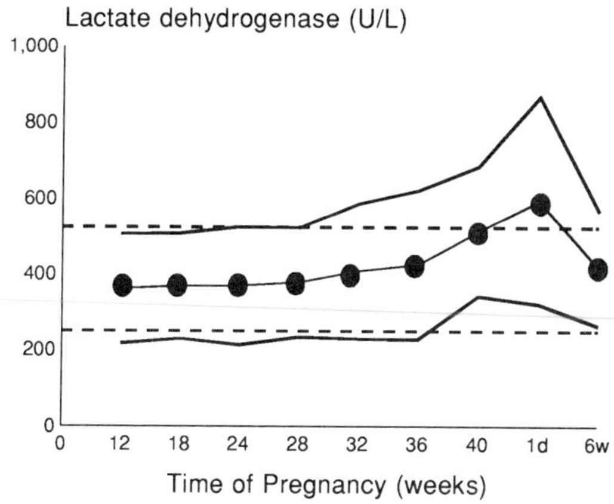

Figure I.E.8. Gestational changes in lactate dehydrogenase activity.

Table I.E.8. Lactate dehydrogenase (U/L)			
Time of Gestation	Reference Range	Mean (sd)	Percentage of Control Mean
Non Pregnant Control	250 - 525	392 (62)	
12 weeks	217 - 506	361 (72)	92
18 weeks	229 - 508	368 (70)	94
24 weeks	213 - 525	369 (78)	94
28 weeks	234 - 523	378 (72)	96
32 weeks	229 - 587	408 (89)	104
36 weeks	227 - 622	425 (99)	108
Term	342 - 686	514 (86)	131
1 day pp	321 - 868	595 (137)	152
6 week pp	264 - 567	416 (76)	106

I.F. Plasma Proteins

EFFECT OF PREGNANCY ON PLASMA PROTEINS

Plasma protein concentrations change markedly during pregnancy. Several physiological factors, including increased plasma volume, increased renal blood flow, altered protein synthetic rates due to hormonal changes, and other acute phase mediators, interact to determine plasma protein concentrations.[1-4] Nutritional deficiencies (protein, zinc, retinol), renal or hepatic disease, infections, immune dysfunctions, and inherited metabolic disorders may also affect protein concentrations[4] (Table I.F.1.). The concentrations of most plasma amino acids are lower in pregnancy than in the non-pregnant state.[5]

Increased renal blood flow and glomerular filtration result in increased renal excretion of both amino acids and protein. Both the quantity and quality of aminoaciduria are altered in pregnancy. As much as 2 g of amino acids may be contained in a 24-hour urine compared with <0.5 g in the non-pregnant state.[5] Glycine, histidine, threonine, serine, and alanine are normally present in the urine of non-pregnant women. Toward term in healthy pregnant women, the urinary concentration of these amino acids was increased by about four times. Other amino acids including lysine, cystine, taurine, phenylalanine, valine, leucine, and tyrosine were excreted in greater amounts in the first trimester but to a lesser extent later in pregnancy.[5] Physiological proteinuria in non-pregnant women is usually characterized by less than 150 mg of protein in the urine per day.[6] Proteinuria may frequently increase to 300 mg/day in pregnancy and should not be considered abnormal until it exceeds 500 mg/day.[7]

Plasma volume expansion causes an overall decrease in plasma protein concentration[1] which is not compensated by the hormonal stimulation of hepatic synthesis of binding and carrier proteins.[8] There is a net decrease in total protein concentration of 10 to 12 g/L, reached by about 28 weeks of gestation.[9,10] This is made up of overall decreases in albumin and the γ-globulin fraction, with increases in the α_1, α_2 and β-globulin fractions. The net albumin:globulin ratio in mid to late pregnancy is 1:1 compared to a non-pregnant ratio of about 1.4:1.

Although the predominant cause of lowered albumin concentration is dilutional, estrogen also tends to reduce albumin concentration. We (see Section III.) and others[2] have shown that mean serum albumin concentration in oral contraceptive users is about 95% of the levels in non-oral contraceptive users.

Despite the net decrease in concentration of total protein, and in particular albumin concentration, when the expanded plasma volume is considered, there is a substantial increase in circulating protein (Table I.F.2.).

This effect is even more marked for the globulin fractions.[11] The estrogen stimulated increase in binding and carrier proteins which make up the globulin fractions is such that their intravascular pool size (protein concentration × plasma volume) is almost double the non-pregnant state.[11] This explains some of the increase in total hormone concentrations. Thus, concentrations of thyroxine and triiodothyronine (bound to thyroxine-binding globulin)[12,13] are greatly increased although the free fractions may be only minimally altered from non-pregnant levels (see Section I.L.). A similar increase occurs in total cortisol (bound to cortisol-binding globulin)[13-15] but an accompanying increase in free cortisol during pregnancy suggests that the higher corticosteroid-binding globulin levels are not the only factor involved in higher cortisol concentrations[15] (see Section I.L.). A similar increase occurs in metal transport proteins such as ceruloplasmin,[14] antiproteases such as α_1-antitrypsin,[16] and proteins required for coagulation and thrombolysis.[17]

TOTAL PROTEIN, ALBUMIN, AND COLLOID OSMOTIC PRESSURE

The effect of expanded plasma volume[18] is apparent in lowered concentrations of total protein and albumin and decreased colloid osmotic pressure which are evident by 12 weeks of gestation and decline further, to stabilize around the 28th week. The postpartum increase in plasma volume which occurs to compensate for intrapartum blood loss results in a further sharp decrease in all three indices on the first postpartum day but, by the 6th postpartum week, all three indices are within the non-pregnant control ranges.

The lower concentration of albumin also results in an apparent decrease in substances normally bound to this protein. Thus, concentrations of total calcium, zinc, and unconjugated bilirubin may be lower than in the non-pregnant state. Concern has been expressed about a lower albumin reserve for bilirubin binding in pregnancy.[19] However, as shown in Table II.F.2., despite a lower concentration, there is actually a net increase in total circulating albumin.

The importance of altered colloid osmotic pressure as a contributing factor to pulmonary and cerebral edema in pregnancy-induced hypertension has stimulated interest in colloid osmotic pressure measurements in pregnancy.[20] Albumin is the major determinant of colloid osmotic pressure and a very strong correlation exists between serum albumin concentration and colloid osmotic pressure measurements.[20] Colloid osmometry is not routinely available in most laboratories but it appears that measurement of colloid osmotic pressure adds little additional information to that obtained from a simple plasma albumin estimation which is usually more easily obtained.

α_1-ANTITRYPSIN

α_1-Antitrypsin is an anti-protease which protects against activity of a variety of proteases including elastase and trypsin.[21-25] Deficiency of α_1-antitrypsin is an important cause of neonatal cholestasis, juvenile cirrhosis, and of premature onset of emphysema and slowly progressive cirrhosis in the adult.[21-25] α_1-Antitrypsin deficiency is caused by inheritance of two of the protease inhibitor (Pi) deficiency alleles, Pi Z, Pi S, and Pi^{null}. Individuals of Pi ZZ, SZ, or Pi^{null} phenotype are at risk for lung disease and the risk is greatly exacerbated by smoking.[21] It has not been clearly proven that heterozygotes with Pi MZ, Pi MS, or homozygous Pi SS phenotypes are at increased risk.[21] Individuals homozygous for the Z phenotype generally have α_1-antitrypsin concentrations of <0.8 g/L. If the mean value for the majority of the population of the MM phenotype is arbitrarily considered to be 100%, the α_1-antitrypsin concentration for a ZZ individual is approximately 16% of the MM mean, while means reported for other phenotypes approximate 75% for MS, 57% for MZ, 52% for SS, and 37% for SZ.[24] When reference ranges are being established, it is essential to remember that absolute values can be modified by factors such as oral contraceptive use (Table 1.F.1. and Section III.).

Indications for α_1-antitrypsin testing include investigation of pulmonary or hepatic disease or a family history of α_1-antitrypsin deficiency. In the latter situation, a couple may be tested for α_1-antitrypsin heterozygosity to determine their risk for having an infant with α_1-antitrypsin deficiency.

It is important to remember that α_1-antitrypsin quantitation may be misleading. Low α_1-antitrypsin concentrations may result from renal or gastrointestinal protein loss or severe liver cirrhosis. Conversely, α_1-antitrypsin is an acute phase reactant and increases in response to estrogen stimulation of the hepatic protein synthesis. Quantitation of α_1-antitrypsin levels therefore cannot be relied on to diagnose α_1-antitrypsin deficiency, although quantitation may indicate the need for phenotyping.[24] Indentification of an individual phenotype by Pi typing is performed by polyacrylamide gel isoelectric focusing.[25]

Unlike individuals who are heterozygous for the Z allele, ZZ individuals do not respond to infection, stress, oral contraceptives, or pregnancy with an increase in α_1-antitrypsin.

α_1-ANTITRYPSIN QUANTITATION AND PROTEASE INHIBITOR TYPING IN ORAL CONTRACEPTIVE USERS

Protease inhibitor typing of our control subjects indicated that 11 were of the MS phenotype, 3 of the MZ phenotype, 1 was ZZ, and the others MM. Of the 11 MS subjects, 3 were oral contraceptive users and 8 were not. Since this was not a random sample, these numbers do not have any implication for frequency of the Z and S alleles in this group, but can be used to illustrate the effect of oral contraceptive use on α_1-antitrypsin quantitation (Table I.F.3). The ZZ individual, previously unidentified as such, was in excellent health and had an α_1-antitrypsin concentration of 0.73 g/L.

α_1-ANTITRYPSIN IN PREGNANCY

Hepatic synthesis of α_1-antitrypsin is strongly influenced by hormonal stimulation. The mean concentration in oral contraceptive users was 135% of the level in non-oral contraceptive users (Section III.).

An early response to estrogen stimulation also occurs in pregnancy. α_1-Antitrypsin concentration was already increased by the 12th week of gestation in our study. Levels increased steadily throughout pregnancy and the mean concentration at term was 191% of the non-pregnant mean concentration. By the 6th postpartum week, α_1-antitrypsin concentration was again within the non-pregnant range.

It is important to be aware that a woman with MZ phenotype, tested in pregnancy by α_1-antitrypsin quantitation alone, may erroneously be considered "normal" because her α_1-antitrypsin concentration is within the reference range. To identify a heterozygote for the Z allele, Pi typing by isoelectric focusing is required.

In our pregnancy group, one woman had α_1-antitrypsin concentrations that showed a typical increase with pregnancy and were within the gestational reference range throughout pregnancy. The specimen obtained at the 6th week postpartum visit however was clearly below the non-pregnant reference range. Phenotyping revealed that she was heterozygous for the Z allele. It is important to emphasize that the gestational increase in α_1-antitrypsin concentration may mask low levels normally found in MZ or SZ phenotypes.

CERULOPLASMIN

Ceruloplasmin is the primary transport protein carrying copper from the liver to peripheral tissues and approximately 80% of copper in plasma is bound to this protein.[26,27] In addition to its role in copper transport, ceruloplasmin is a major serum antioxidant, aids in transfer of iron to transferrin, participates in the inflammatory response, promotes growth in cell cultures, stimulates angiogenesis in corneas and cell cultures, and regulates synthesis of biogenic amines.[26]

Several factors increase the concentration of ceruloplasmin in human serum or plasma. Ceruloplasmin is much higher in smokers than in non-smoking women.[28] Ceruloplasmin increases moderately with the acute phase response.[29] Plasma ceruloplasmin concentration responds strongly to the hormonal effect of pregnancy and oral contraceptives.[13,14,30,31]

The mean ceruloplasmin concentration in our oral contraceptive users was 164% of the mean in the non-oral contraceptive users (Section III.). The mean ceruloplasmin concentration at term was 210% of the non-pregnant mean. The importance of ceruloplasmin in normal mammalian cell growth and development is shown in the twofold increase in plasma levels of this protein in situations of estrogen-induced hepatic protein synthesis.

CERULOPLASMIN IN GENETIC DISORDERS OF COPPER METABOLISM

In approximately 90% of patients with Wilson's disease, ceruloplasmin is extremely low, well below the reference range for non-pregnant individuals.[32] However, in 10% of individuals with Wilson's disease, although low, ceruloplasmin may be within the adult reference range.[32] Calculation of non-ceruloplasmin bound copper may be helpful in the diagnosis of Wilson's disease in the latter circumstances.[33] Unbound copper (μmol/L), calculated as [total copper (μmol/L) − {ceruloplasmin (mg/L) × 0.047}] greater than 10 μmol/L, suggests Wilson's disease.[33]

In women with Wilson's disease who become pregnant, ceruloplasmin concentrations together with serum copper concentrations increase with progressive gestation.[30,31]

COMPLEMENT PROTEINS, C3 AND C4

Plasma concentrations of the complement fractions C3 and C4 vary in the magnitude of change with increasing gestation. We noted that C3 was already increased by 12 weeks and continued to increase to term. In contrast, C4 remained essentially unchanged from the non-pregnant control range and only increased slightly between 36 weeks and term. Others have found the mean values for C3 and C4 at term to be 135 and 124% of the non-pregnant control mean values.[34]

Reference ranges for complement assays vary greatly between laboratories.[35] In one multicenter study involving 14 laboratories, lower limits of reference ranges for C3 ranged from 0.7 to 1.1 g/L, the upper limit from 1.5 to 2.3 g/L, and the interval between upper and lower limits varied widely. Similarly lower limits for C4 ranged from 0.10 to 0.28 and upper limits from 0.35 to 0.80 g/L, although the range interval was more consistent between laboratories.[35]

USE IN PREGNANCY

Hypocomplementemia is an important marker for immune-complex-mediated disease and serial studies are used to assess disease activity. Systemic lupus erythematosis (SLE), which affects women of child bearing age, may adversely affect the fetus (prematurity, heart block, intrauterine death) and mother (hypertension and proteinuria).[36] The decrease in C3 and C4 with increased disease activity can be used to differentiate hypertension and proteinuria of SLE from pregnancy-induced hypertension where complement levels are similar to that of normal pregnancy.[34,36] However, the pregnancy-induced increase in C3 may obscure the effect of disease such as SLE so that a "normal" complement level in a pregnant woman with SLE may be found despite disease activity.[36,37] It is therefore especially important that measurements of C3 levels in pregnancy be compared to gestation-specific reference ranges established by the same laboratory.

OTHER CONDITIONS THAT CAN ALTER COMPLEMENT LEVELS

Complement levels may increase in bacterial infection[38] and in chronic non-infectious inflammatory states such as rheumatoid arthritis.[35] Severe sepsis however can be associated with decreased levels of both C3 and C4.[38] Severe malnutrition decreases C3 but not C4 levels[39] while severe liver failure decreases both.[40] Complement levels may be altered by inherited abnormalities of complement pathway components or one of the complement control proteins.[41] Between 20 and 40% of the normal population have one or two C4 "null" genes with correspondingly low C4 levels.[35]

IMMUNOGLOBULINS IgG, IgA, AND IgM

Transplacental transfer of IgG is facilitated by the presence of placental receptors specific for the Fc portion of IgG, while passage of IgM and IgA is blocked.[42] The transfer of all IgG subclasses, particularly IgG_1 and IgG_3,[43] leads to a decrease in maternal IgG plasma half-life and IgG concentrations decline, reaching a nadir in the third trimester.

We found that IgG concentration was already decreased by week 12 and declined steadily until term. Values were within the non-pregnant range by 6 weeks postpartum. Concentrations of both IgA and IgM were slightly reduced relative to the non-pregnant reference range throughout pregnancy returning to non-pregnant levels by the 6th postpartum week.

Many diseases associated with autoimmunity or immunodeficiency have a peak incidence during the childbearing years and pregnancy may occur in women with combined variable immunodeficiency or with selective IgA deficiency.[42]

Combined variable immunodeficiency, with humoral deficiency of some or all immunoglobulin classes, is characterized by tendency to recurrent bacterial infections, low immunoglobulin levels, and depressed antibody response.[44] Disorders associated with this condition include rheumatoid arthritis and systemic lupus erythematosis, as well as autoimmune phenomena such as hemolytic anemia, pernicious anemia, and immune thrombocytopenia.[42] Hypogammaglobulinemic women have given birth to infants with normal immunologic maturation.[45] However, if severe deficiency occurs during pregnancy, immunoglobulin replacement therapy may be given intravenously or subcutaneously.[46]

Selective IgA deficiency is very common (prevalence 1:400 to 1:800).[47] Frequent respiratory infections and chronic diarrhea result from deficiency of secretory IgA. There is also an increased incidence of allergic disorders such as asthma and autoimmune disorders such as celiac disease or pernicious anemia.[42] Asymptomatic women with selective IgA deficiency who become pregnant should be identified as such because IgA in plasma, whole blood, or γ-globulin infusions may precipitate an anti-IgA response with anaphylaxis.[42]

RETINOL-BINDING PROTEIN AND PREALBUMIN

Retinol-binding protein (RBP) is the vitamin A (retinol) carrier protein transporting retinol from the liver to peripheral tissues. Apo-RBP is synthesized in the hepatocyte and combined with *all-trans*-retinol in a 1:1 molecular complex to form RBP. On release into plasma, RBP binds with prealbumin (transthyretin) which also binds a single molecule of thyroxine.[48,49] RBP binds to specific cell surface receptors, releasing retinol into peripheral tissue cells. The apo-RBP is released in an altered conformation which no longer binds to prealbumin and is catabolized by the kidney.[48,49]

RBP has a very short plasma half-life of about 12 hours and has been proposed as an index of acute protein depletion.[4] Low values are seen in vitamin A deficiency, zinc deficiency, hyperthyroidism, and liver disease.[4]

Prealbumin is also synthesized in the liver and, with a plasma half-life of 2 days, is probably the most useful of the rapid turnover proteins, as an index of protein depletion.[4] Prealbumin transports approximately one third of protein-bound thyroxine.[4] The complex of prealbumin with RBP appears to stabilize retinol binding and protect the low molecular weight RBP from renal catabolism. Prealbumin is a negative acute phase reactant and concentrations also decrease if hepatic protein synthesis is decreased.[4,49]

There are genetic variations in prealbumin,[49] one of which is characterized by increased thyroxine affinity.[50] A euthyroid man had hyperthyroxinemia due to increased binding of thyroxine to the variant prealbumin.[50] Other similar cases have since been described.[49] Other genetic variants have been implicated in familial amyloidotic polyneuropathy, an autosomal dominant disorder, apparently related to mutations in the prealbumin gene on chromosome 18.[51]

RETINOL-BINDING PROTEIN AND PREALBUMIN IN PREGNANCY

Although both RBP and prealbumin behave as negative acute phase reactants, decreasing with acute stress,[4] they appear to increase when hepatic protein synthesis is stimulated by increased estrogen. Significantly higher concentrations of both proteins were found in our oral contraceptive users than in our controls who were not using oral contraceptives (Section III.).

The plasma concentrations during pregnancy reflect the interaction between the dilutional effect of plasma volume and stimulation of hepatic synthesis by estrogen, with vitamin A nutritional status and alteration in other plasma thyroid-binding hormones as additional modifying factors.

RBP was marginally decreased at 12 weeks and thereafter remained within the limits of the control reference range. Prealbumin levels were generally moderately lower than in the non-pregnant state.

TRANSFERRIN

Transferrin is the major transport protein for iron and transferrin receptors have been identified on placental villous epithelial cells.[52] Transferrin is present in the yolk sac of the early embryo.[52] Transferrin has been shown to be an important growth factor stimulating *in vitro* cellular proliferation.[53]

Transferrin is clinically important in several contexts. As a "rapid turnover" protein with a plasma half-life of approximately 7 days, it has been suggested as a more sensitive indicator of nutritional protein depletion than albumin concentration.[4] However, the transferrin concentration and iron-binding capacity in plasma is also altered by the iron status of the individual, stimulation by estrogen, and by acute phase mediators.

Transferrin is a negative acute phase reactant[4,54] decreasing during the stress response. Unlike the increase in positive acute phase reactants which occurs about 8 hours after acute injury, the decrease in acute phase reactants such as albumin and transferrin occurs almost immediately.[54] This may partially account for the rapid drop in transferrin concentration within 24 hours of delivery.

The indices of transferrin concentration and iron-binding capacity increase in iron deficiency[4] while transferrin saturation decreases. Estrogen stimulation is important in increasing plasma transferrin concentration of pregnancy but decreasing iron stores may also be important in this situation. The effect of oral contraceptive use on the non-pregnant reference range for transferrin was less marked than that seen for ceruloplasmin (Table III.B.3.).

Other factors can influence transferrin concentration. Altered plasma transferrin concentrations have been reported in several endocrine disorders.[55] Transferrin was extremely elevated in acromegaly, and decreased in hypopituitarism, while moderately lower in some patients with Cushing's syndrome and hyperthyroidism.[55]

TRANSFERRIN IN PREGNANCY (SEE ALSO SECTION I.K INDICES OF IRON METABOLISM)

We found that the mean and range of transferrin concentration at 12 weeks was similar to the non-pregnant female control values. A steady increase was seen from 18 weeks to term when the pregnancy group mean was 170% of the non-pregnant mean. A significant decrease in transferrin concentration was already noted by 1 day postpartum and, by the 6th postpartum week, transferrin levels had returned to control values.

REFERENCES

1. **Goodlin, R. C., Dobry, C. A., Anderson, J. C., Woods, R. E., and Quaife, M.,** Clinical signs of normal plasma volume expansion during pregnancy, *Am. J. Obstet. Gynecol.*, 145, 1001, 1983.
2. **Knopp, R. H., Bergelin, R. O., Wahl, P. W., Walden, C. E., and Chapman, M. B.,** Clinical chemistry alterations in pregnancy and oral contraceptive use, *Obstet. Gynecol.*, 66, 682, 1985.
3. **Heinrich, P. C., Castell, J. V., and Andus, T.,** Interleukin-6 and the acute phase response, *Biochem. J.*, 265, 621, 1990.
4. **Benjamin, D. R.,** Laboratory tests and nutritional assessment: protein-energy status, *Pediatr. Clin. N.A.*, 36, 1391, 1989.
5. **Hytten, F. E. and Cheyne, G. A.,** The aminoaciduria of pregnancy, *J. Obstet. Gynaecol. Br. Commonw.*, 79, 424, 1972.
6. **Abuelo, J. G.,** Proteinuria: diagnostic principles and procedures, *Ann. Intern. Med.*, 98, 186, 1983.
7. **Davidson, J. M.,** The effect of pregnancy on kidney function in renal allograft recipients, *Kidney Int.*, 27, 74, 1985.
8. **Van Thiel, D. H. and Gavaler, J. S.,** Pregnancy associated steroids and their effects on the liver, *Sem. Liver Dis.*, 7, 1, 1987.
9. **Lind, T.,** Clinical chemistry of pregnancy, *Adv. Clin. Chem.*, 21, 1, 1980.
10. **National Research Council, Committee on Nutrition of the Mother and Preschool Child,** *Laboratory Indices of Nutritional Status in Pregnancy,* National Academy of Sciences, Washington, DC, 1978, 1.
11. **Rosso, P.,** *Nutrition and Metabolism in Pregnancy,* Oxford University Press, New York, 1990, 1.
12. **Becks, G. P. and Burrow, G. N.,** Thyroid disease and pregnancy, *Med. Clin. N.A.*, 75, 121, 1991.
13. **Sobbrio, G. A., Granata, A., Granese, D., D'Arrigo, F., Panacea, A., Nicita, R., Pulle, C., and Trimarchi, F.,** Sex hormone binding globulin, cortisol binding globulin, thyroxine binding globulin, ceruloplasmin: changes in treatment with two oral contraceptives low in oestrogen, *Clin. Exp. Obstet. Gynecol.*, 18, 43, 1991.
14. **Song, S., Chen, J.-K., He, M.-L., and Fotherby, K.,** Effect of some oral contraceptives on serum concentrations of sex hormone binding globulin and ceruloplasmin, *Contraception,* 39, 385, 1989.
15. **Scott, E. M., McGarrigle, H. H. G., and Lachelin, G. C. L.,** The increase in plasma and saliva cortisol levels in pregnancy is not due to the increase in corticosteroid binding globulin levels, *J. Clin. Endocrinol. Metab.*, 71, 639, 1990.
16. **Barr, A. E., Cole, D. A., and Clague, A. E.,** The effect of an acute phase response or pregnancy on plasma alpha-1-antitrypsin concentrations in persons with various S and Z phenotypes, *Ann. Clin. Biochem.*, 25, 255, 1988.
17. **Beller, F. K. and Ebert, C.,** Effects of oral contraceptives on blood coagulation, *Obstet. Gyn. Surv.*, 40, 425, 1985.
18. **Hytten, F. E.,** Blood volume changes in normal pregnancy, *Clin. Haematol.*, 14, 601, 1985.
19. **Esbjorner, E., Jarnerot, G., Sanstrom, B., and Ostling, G.,** Serum albumin reserve for bilirubin binding during pregnancy in healthy women, *Obstet. Gynecol.*, 73, 93, 1989.
20. **Nguyen, H. N., Clark, S. L., Greenspoon, J., Diesfield, P., and Wu, P. Y. K.,** Peripartum colloid osmotic pressure: correlation with serum proteins, *Obstet. Gynecol.*, 68, 807, 1986.
21. **Feld, R. D.,** Heterozygosity of α_1-antitrypsin: a health risk?, *Crit. Rev. Clin. Lab. Sci.*, 27, 461, 1989.

22. **Pierce, J. A.,** Antitrypsin and emphysema, *JAMA*, 259, 2890, 1988.
23. **Eriksson, S.,** α_1-Antitrypsin and liver cirrhosis in adults. An analysis of 35 Swedish autopsied cases, *Acta Med. Scand.*, 221, 461, 1987.
24. **Lieberman, J., Gaidulis, L., Garoutte, B., and Mittman, C.,** Identification and characteristics of the common alpha$_1$-antitrypsin phenotypes, *Chest*, 62, 557, 1972.
25. **Fagerhol, M. K. and Cox, D. W.,** The Pi polymorphism: genetic, biochemical and clinical aspects of human alpha$_1$-antitrypsin, *Adv. Hum. Genet.*, 11, 1, 1981.
26. **Frieden, E.,** Perspectives on copper biochemistry, *Clin. Physiol. Biochem.*, 4, 11, 1986.
27. **Harris, E. D.,** Copper transport: an overview, *Proc. Soc. Exp. Biol.*, 196, 130, 1991.
28. **Galdston, M., Feldman, J. G., and Levytska, V.,** Antioxidant activity of serum ceruloplasmin and transferrin available iron-binding capacity in smokers and non-smokers, *Am. Rev. Resp. Dis.*, 135, 783, 1987.
29. **Fleck, A.,** Clinical and nutritional aspects of changes in acute-phase proteins during inflammation, *Proc. Nutr. Soc.*, 48, 347, 1989.
30. **Walshe, J. M.,** Pregnancy in Wilson's disease, *Q. J. Med.*, XLVI, 73, 1977.
31. **Dupont, P., Irion, O., and Beguin, F. P.,** Pregnancy in a patient with treated Wilson's disease: a case report, *Am. J. Obstet. Gynecol.*, 163, 1527, 1990.
32. **Brewer, G. J. and Yuzbasiyan-Gurkan, V.,** Wilson's disease: an update with emphasis on new approaches to treatment, *Dig. Dis.*, 7, 178, 1989.
33. **Gambino, R.,** Free serum copper — best test for Wilson's disease, *Lab. Rep.*, 14, 1, 1992.
34. **Gallery, E. D., Raffos, J., Gyory, A. Z., and Wells, J. V.,** A prospective study of serum complement (C3 and C4) levels in normal human pregnancy: effect of the development of pregnancy-associated hypertension, *Aust. N.Z. J. Med.*, 11, 243, 1981.
35. **Hebert, L. A., Cosio, F. G., and Neff, J. C.,** Diagnostic significance of hypocomplementemia, *Kidney Int.*, 39, 811, 1991.
36. **Buyon, J. P., Cronstein, B. N., Morris, M., Tanner, M., and Weissmann, G.,** Serum complement values (C3 and C4) to differentiate between systemic lupus activity and pre-eclampsia, *Am. J. Med.*, 81, 194, 1986.
37. **Schena, F. P., Manno, C., Selvaggi, L., Loverro, G., Betocchi, S., and Bonomo, L.,** Behaviour of immune complexes and the complement system in normal pregnancy and pre-eclampsia, *J. Clin. Lab. Immunol.*, 7, 21, 1982.
38. **Leon, C., Rodrigo, M. J., Tomasa, A., Gallart, M. T., Latorre, F. J., Rius, J., and Brugues, J.,** Complement activation in septic shock due to gram-negative and gram-positive bacteria, *Crit. Care Med.*, 10, 308, 1982.
39. **Haller, L., Zubler, R. H., and Lambert, P. H.,** Plasma levels of complement components and complement haemolytic activity in protein-energy malnutrition, *Clin. Exp. Immunol.*, 34, 248, 1978.
40. **Wyke, R. J., Rajkovic, I. A., and Williams, R.,** Impaired opsonization by serum from patients with chronic liver diseases, *Clin. Exp. Immunol.*, 51, 91, 1983.
41. **Schur, P. H.,** Inherited complement component abnormalities, *Annu. Rev. Med.*, 37, 333, 1986.
42. **Levy, Y. and Douglas, S.,** Immunologic disorders, in *Principles of Medical Therapy in Pregnancy*, Gleicher, N., Ed., Plenum Press, New York, 1985, 948.
43. **Pitcher-Wimott, R. W., Hindocha, P., and Wood, C. B. S.,** The placental transfer of IgG subclasses in human pregnancy, *Clin. Exp. Immunol.*, 41, 303, 1980.
44. **Hermans, P. E., Diaz-Buxo, J. A., and Stobo, J. D.,** Idiopathic late-onset immunoglobulin deficiency: clinical observations in 50 patients, *Am. J. Med.*, 61, 221, 1976.
45. **Kobayashi, R. H., Hyman, C. J., and Stiehm, E. R.,** Immunologic maturation in an infant born to a mother with agammaglobulinemia, *Am. J. Dis. Child.*, 134, 942, 1980.
46. **Berger, M., Cupps, T. R., and Fauci, A. S.,** High-dose immunoglobulin replacement by slow subcutaneous infusion during pregnancy, *JAMA*, 247, 2824, 1982.

47. **Ammann, A. J. and Hong, R.,** Selective IgA deficiency: presentation of 30 cases and review of the literature, *Medicine,* 50, 223, 1971.
48. **Olson, J. A.,** Recommended dietary intakes of vitamin A in humans, *Am. J. Clin. Nutr.,* 45, 704, 1987.
49. **Bartalena, L.,** Recent achievements in studies on thyroid hormone-binding proteins, *Endocr. Rev.,* 11, 47, 1990.
50. **Moses, A. C., Lawlor, J., Haddow, J., and Jackson, I. M. D.,** Familial euthyroid hyperthyroxinemia resulting from increased thyroxine binding to thyroxine-binding prealbumin, *N. Engl. J. Med.,* 306, 966, 1982.
51. **Sakaki, Y., Yoshioka, K., Tanahashi, H., Furuya, H., and Sasaki, H.,** Human transthyretin (prealbumin) gene and molecular genetics of familial amyloidotic polyneuropathy, *Mol. Biol. Med.,* 6, 161, 1989.
52. **Okuyama, T., Tawada, T., Furuya, H., and Villee, C. A.,** The role of transferrin and ferritin in the fetal-maternal placental unit, *Am. J. Obstet. Gynecol.,* 152, 344, 1985.
53. **Hayashi, I. and Sato, G. H.,** Replacement of serum by hormone permits growth of cells in a defined medium, *Nature,* 259, 123, 1976.
54. **Fleck, A.,** Clinical and nutritional aspects of changes in acute phase proteins during inflammation, *Proc. Nutr. Soc.,* 48, 347, 1989.
55. **Matsubara, M., Koyanagawa, Y., Odagaki, E., and Nakagawa, K.,** Plasma transferrin in abnormal endocrine states, *Horm. Metab. Res.,* 21, 335, 1989.

I.F.1. FACTORS WHICH AFFECT SERUM OR PLASMA PROTEIN CONCENTRATIONS

Table I.F.1. Factors which affect serum or plasma protein concentrations

Region	Protein	APR[a]	Hormonal (OC) Effect	Other Factors
Pre-albumin	Prealbumin	↓	↑	↓ zinc, protein deficiency
	RBP		↑	↑ retinol excess ↓ zinc, protein retinol deficiency
Albumin	Albumin	↓	↓	↓ due to plasma volume increase, also renal, liver disease and nutritional protein deficiency
α-1 region	α_1-antitrypsin	↑↑↑	↑↑↑	↓ in Z and S genetic variants
α-2-region	ceruloplasmin	↑↑↑	↑↑↑	↓ Wilson's disease, Cu deficiency. ↑ in cholestasis, smoking
ß-1-region	transferrin	↓↓	↑↑	↑ iron deficiency ↓ in hemochromatosis
	Complement C4	↑		↓ classic pathway activation ↓ severe liver disease, severe sepsis, burns
ß-2-region	fibrinogen	↑	↑↑	↓ in hypercoagulable st ates
	Complement C3		↑	↓ alternative or classic pathway activation ↓ severe liver disease, sepsis or malnutrition
gamma-region	Immunoglobulins IgG, IgA, IgM			↓ in immunodeficiency states ↑ by infection
	CRP[b]	↑↑↑		↑ acutely in sepsis, trauma

[a] Acute phase reactant [b] C-reactive protein

I.F.2. COMPARISON OF PLASMA CONCENTRATION AND TOTAL CIRCULATING PLASMA PROTEINS IN NON-PREGNANT STATE AND AT 28 WEEKS GESTATION USING MEAN PLASMA VOLUME ESTIMATIONS FROM HYTTEN[18]

Table I.F.2. Comparison of plasma concentration and total circulating plasma proteins in non-pregnant state and at 28 weeks gestation using mean plasma volume estimations from Hytten.[18]

	non-pregnant state mean plasma volume 2600 mL		28 weeks gestation mean plasma volume 3850 mL	
Protein	concentration (g/L)	total (g)	concentration (g/L)	total (g) [% increase]
Total protein	71	185	59	227 [123]
Albumin	41	107	32	123 [115]
α_1-Antitrypsin	2.15	5.59	3.62	13.94 [250]
Ceruloplasmin	252	655	477	1836 [280]
Thyroxine-binding globulin	22	57	39	150 [263]
Transferrin	2.85	7.41	3.62	13.94 [188]

I.F.3. EFFECT OF ORAL CONTRACEPTIVE USE ON MEAN (SD) OF α_1-ANTITRYPSIN CONCENTRATION (G/L)

Table I.F.3. Effect of oral contraceptive use on mean (sd) of α_1-antitrypsin concentration (g/L).

Phenotype	No oral contraceptive use	Oral contraceptive use
M M	2.2 (0.46)	3.1 (0.85)
M S	1.99 (0.35)	3.05 (1.05) - only 3 subjects (values 4.53, 2.53, 2.73,)
M Z	1.37 (0.16)	No subjects

I.F.4. TOTAL PROTEIN

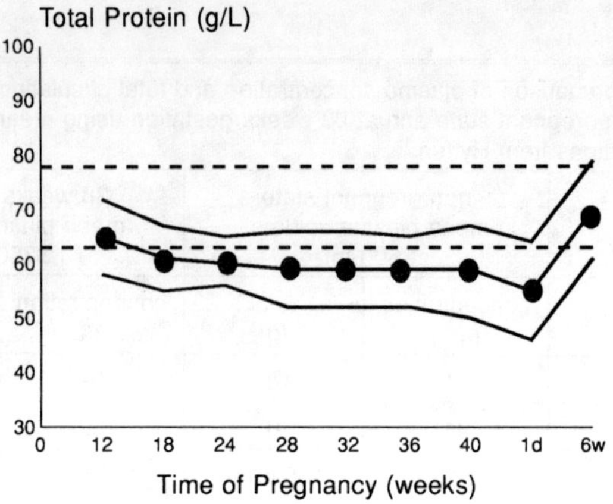

Figure I.F.4. Gestational changes in total protein concentration.

Table I.F.4. Total protein (g/L)

Time of Gestation	Reference Range	Mean (sd)	Percentage of Control mean
Non Pregnant Control	63 - 78	71 (4)	
12 weeks	58 - 72	65 (3)	92
18 weeks	55 - 67	61 (3)	86
24 weeks	56 - 64	60 (2)	85
28 weeks	52 - 66	59 (3)	83
32 weeks	52 - 66	59 (3)	83
36 weeks	52 - 65	59 (3)	83
Term	50 - 67	59 (4)	83
1 day pp	46 - 64	55 (4)	77
6 week pp	61 - 79	70 (4)	99

I.F.5. ALBUMIN

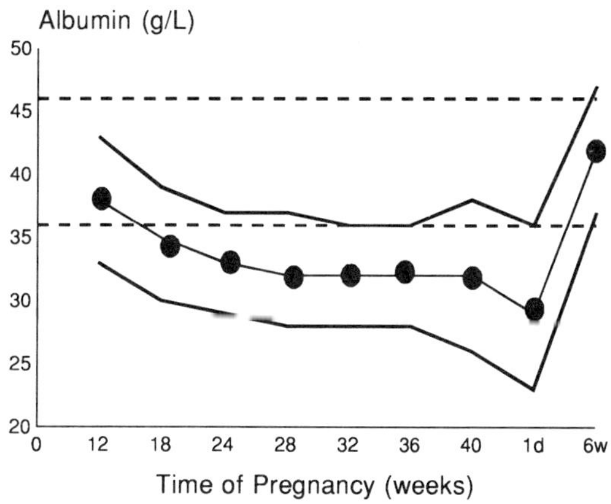

Figure I.F.5. Gestational changes in albumin concentration.

Table I.F.5. Albumin (g/L)			
Time of Gestation	Reference Range	Mean (sd)	Percentage of Control mean
Non Pregnant Control	36 - 46	41 (2)	
12 weeks	33 - 43	38 (2)	93
18 weeks	30 - 39	35 (2)	85
24 weeks	29 - 37	33 (2)	80
28 weeks	28 - 37	32 (2)	78
32 weeks	28 - 36	32 (2)	78
36 weeks	28 - 36	32 (2)	78
Term	26 - 38	32 (3)	78
1 day pp	23 - 36	29 (3)	71
6 week pp	37 - 47	42 (3)	102

I.F.6. COLLOID OSMOTIC PRESSURE

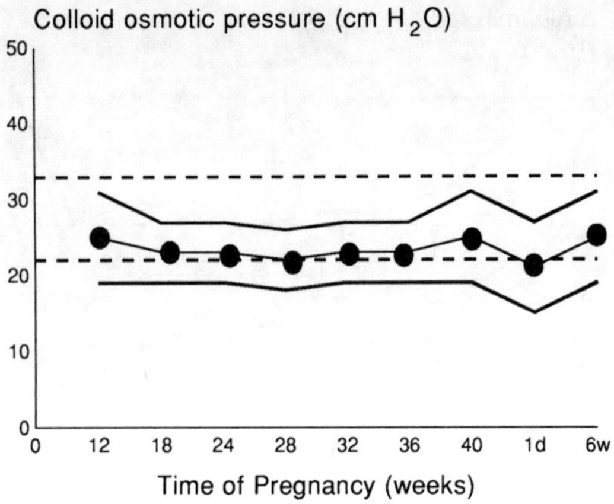

Figure I.F.6. Gestational changes in colloid osmotic pressure.

Table I.F.6. Colloid osmotic pressure (cm H_2O)

Time of Gestation	Reference Range	Mean (sd)	Percentage of Control mean
Non Pregnant Control	22 - 33	27 (2)	
12 weeks	19 - 31	25 (3)	93
18 weeks	19 - 27	23 (2)	85
24 weeks	19 - 27	23 (2)	85
28 weeks	18 - 26	22 (2)	81
32 weeks	19 - 27	23 (2)	85
36 weeks	19 - 27	23 (2)	85
Term	19 - 31	25 (3)	93
1 day pp	15 - 27	21 (3)	78
6 week pp	19 - 31	25 (3)	93

I.F.7. ALPHA-1-ANTITRYPSIN

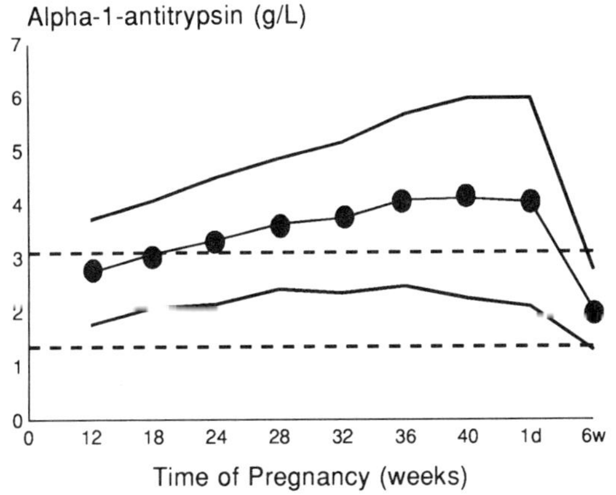

Figure I.F.7. Gestational changes in alpha-1-antitrypsin concentration

Table I.F.7. α-1-Antitrypsin (g/L)			
Time of Gestation	Reference Range	Mean (sd)	Percentage of Control mean
Non Pregnant Control	1.35 - 3.10 #	2.15 (0.42)	
12 weeks	1.76 - 3.72	2.74 (0.49)	129
18 weeks	2.07 - 4.07	3.07 (0.50)	143
24 weeks	2.14 - 4.50	3.32 (0.59)	154
28 weeks	2.41 - 4.85	3.62 (0.61)	169
32 weeks	2.34 - 5.14	3.74 (0.70)	174
36 weeks	2.47 - 5.67	4.07 (0.80)	189
Term	2.24 - 5.96	4.10 (0.93)	191
1 day pp	2.09 - 5.97	4.03 (0.97)	187
6 week pp	1.28 - 2.76	2.02 (0.37)	93

Non-pregnant controls - data from 1 ZZ subject and OC users excluded, but 3 MZ subjects not excluded.
Pregnant women - data from 1 MZ subject not excluded.

I.F.8. CERULOPLASMIN

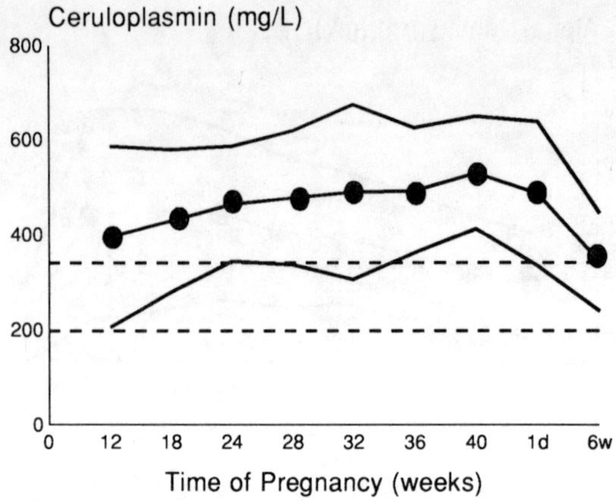

Figure I.F.8. Gestational changes in ceruloplasmin concentration.

Table I.F.8. Ceruloplasmin (mg/L)

Time of Gestation	Reference Range	Mean (sd)	Percentage of Control mean
Non Pregnant Control	198 - 340 #	252 (38)	
12 weeks	205 - 586	395 (95)	157
18 weeks	279 - 579	429 (75)	170
24 weeks	342 - 586	464 (61)	184
28 weeks	335 - 618	477 (71)	189
32 weeks	305 - 673	489 (92)	194
36 weeks	360 - 624	492 (66)	195
Term	411 - 648	529 (59)	210
1 day pp	333 - 637	485 (76)	192
6 week pp	239 - 444	341 (51)	135

Non-pregnant control reference range - data from OC users excluded

I.F.9. COMPLEMENT C3

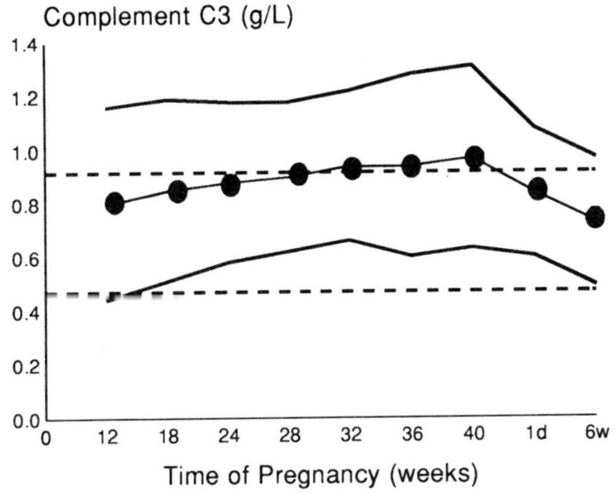

Figure I.F.9. Gestational changes in complement C3

Time of Gestation	Reference Range	Mean (sd)	Percentage of Control mean
Non Pregnant Control	0.47 - 0.92 #	0.71 (0.12)	
12 weeks	0.44 - 1.16	0.80 (0.18)	112
18 weeks	0.51 - 1.19	0.85 (0.17)	120
24 weeks	0.58 - 1.18	0.88 (0.15)	124
28 weeks	0.62 - 1.18	0.90 (0.14)	127
32 weeks	0.66 - 1.22	0.94 (0.14)	132
36 weeks	0.60 - 1.26	0.94 (0.17)	132
Term	0.64 - 1.31	0.98 (0.17)	138
1 day pp	0.60 - 1.08	0.84 (0.12)	118
6 week pp	0.49 - 0.97	0.73 (0.12)	103

Table I.F.9. Complement C3 (g/L)

Non-pregnant control reference range - data from OC users excluded

I.F.10. COMPLEMENT C4

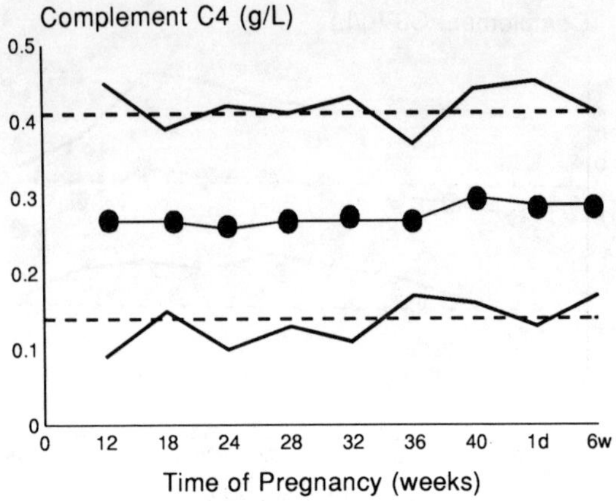

Figure I.F.10. Gestational changes in complement C4 concentration.

Table I.F.10. Complement C4 (g/L)			
Time of Gestation	Reference Range	Mean (sd)	Percentage of Control mean
Non Pregnant Control	0.14 - 0.41	0.25 (0.07)	
12 weeks	0.09 - 0.45	0.27 (0.09)	108
18 weeks	0.15 - 0.39	0.27 (0.06)	108
24 weeks	0.10 - 0.42	0.26 (0.08)	104
28 weeks	0.13 - 0.41	0.27 (0.07)	108
32 weeks	0.11 - 0.43	0.27 (0.08)	108
36 weeks	0.17 - 0.37	0.27 (0.05)	108
Term	0.16 - 0.44	0.30 (0.07)	120
1 day pp	0.13 - 0.45	0.29 (0.08)	116
6 week pp	0.17 - 0.41	0.29 (0.06)	116

I.F.11. IMMUNOGLOBULIN A

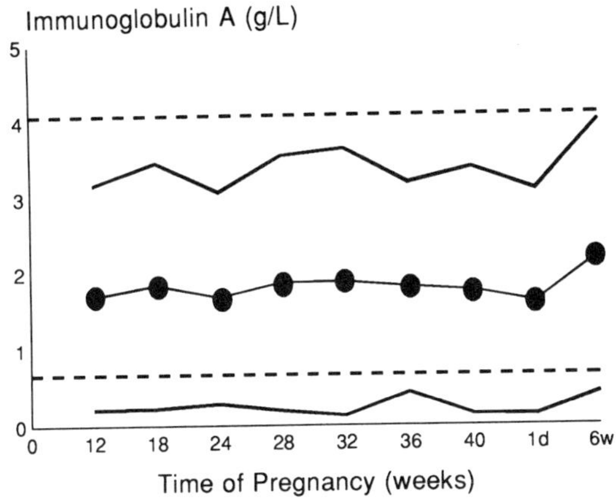

Figure I.F.11. Gestational changes in immunoglobulin A concentration.

Table I.F.11. Immunoglobulin A (g/L)

Time of Gestation	Reference Range	Mean (sd)	Percentage of Control mean
Non Pregnant Control	0.68 - 4.08	2.22 (0.84)	
12 weeks	0.21 - 3.17	1.69 (0.74)	76
18 weeks	0.23 - 3.43	1.83 (0.80)	82
24 weeks	0.30 - 3.06	1.68 (0.69)	76
28 weeks	0.19 - 3.55	1.87 (0.84)	84
32 weeks	0.12 - 3.64	1.88 (0.88)	85
36 weeks	0.43 - 3.19	1.81 (0.69)	82
Term	0.14 - 3.38	1.76 (0.81)	79
1 day pp	0.13 - 3.09	1.61 (0.74)	73
6 week pp	0.45 - 3.97	2.21 (0.88)	100

I.F.12. IMMUNOGLOBULIN G

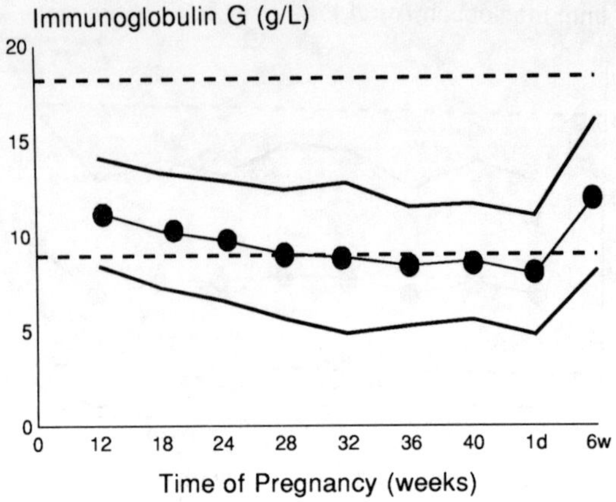

Figure I.F.12. Gestational changes in immunglobulin G concentration.

Table I.F.12. Immunoglobulin G (g/L)

Time of Gestation	Reference Range	Mean (sd)	Percentage of Control mean
Non Pregnant Control	8.93 - 18.21	13.20 (2.28)	
12 weeks	8.38 - 14.10	11.24 (1.43)	85
18 weeks	7.22 - 13.30	10.26 (1.52)	78
24 weeks	6.54 - 12.90	9.72 (1.59)	74
28 weeks	5.58 - 12.38	8.98 (1.70)	68
32 weeks	4.81 - 12.73	8.77 (1.98)	66
36 weeks	5.22 - 11.46	8.34 (1.56)	63
Term	5.54 - 11.62	8.58 (1.52)	65
1 day pp	4.76 - 11.00	7.88 (1.56)	60
6 week pp	8.13 - 16.01	12.07 (1.97)	91

I.F.13. IMMUNOGLOBULIN M

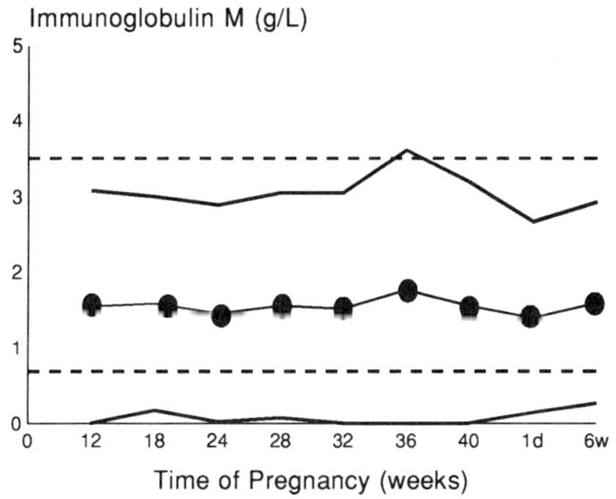

Figure I.F.13. Gestational changes in immunoglobulin M concentration.

Table I.F.13. Immunoglobulin M (g/L)			
Time of Gestation	Reference Range	Mean (sd)	Percentage of Control mean
Non Pregnant Control	0.69 - 3.51	1.94 (0.74)	
12 weeks	0.01 - 3.09	1.55 (0.77)	80
18 weeks	0.17 - 3.01	1.59 (0.71)	82
24 weeks	0.02 - 2.90	1.46 (0.72)	75
28 weeks	0.07 - 3.06	1.56 (0.75)	80
32 weeks	0 - 3.06	1.52 (0.77)	78
36 weeks	0 - 3.61	1.77 (0.92)	91
Term	0 - 3.20	1.56 (0.82)	80
1 day pp	0.12 - 2.68	1.40 (0.64)	72
6 week pp	0.26 - 2.94	1.60 (0.67)	82

I.F.14. PREALBUMIN

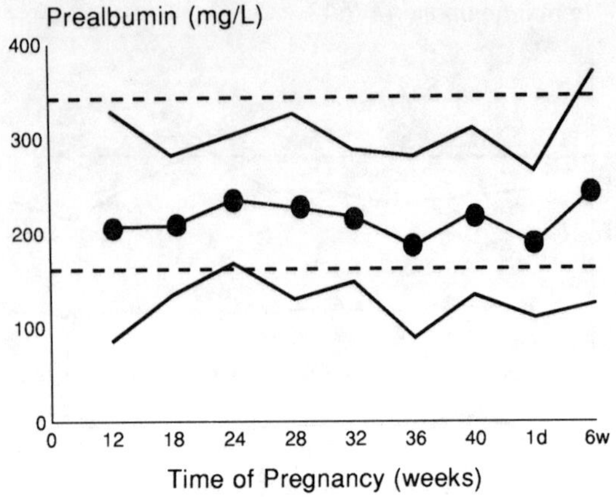

Figure I.F.14. Gestational changes in prealbumin concentration.

Table I.F.14. Prealbumin (mg/L)			
Time of Gestation	Reference Range	Mean (sd)	Percentage of Control mean
Non Pregnant Control	161 - 342 #	246 (46)	
12 weeks	85 - 329	207 (61)	84
18 weeks	134 - 282	208 (37)	85
24 weeks	167 - 303	235 (32)	96
28 weeks	129 - 325	227 (49)	91
32 weeks	147 - 287	217 (35)	88
36 weeks	88 - 280	184 (48)	75
Term	133 - 309	221 (44)	90
1 day pp	109 - 265	187 (39)	76
6 week pp	124 - 368	246 (61)	100

Non-pregnant control reference range - data from OC users excluded

I.F.15. RETINOL-BINDING PROTEIN

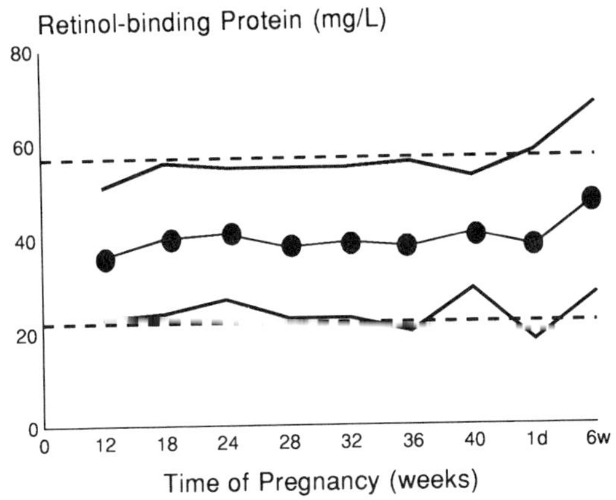

Figure I.F.15. Gestational changes in retinol-binding protein concentration.

Table I.F.15. Retinol-binding protein (mg/L)

Time of Gestation	Reference Range	Mean (sd)	Percentage of Control mean
Non Pregnant Control	22 - 57 #	38 (8)	
12 weeks	23 - 51	36 (7)	95
18 weeks	24 - 56	40 (8)	105
24 weeks	27 - 55	41 (7)	108
28 weeks	23 - 55	38 (8)	100
32 weeks	23 - 55	39 (8)	103
36 weeks	20 - 56	38 (9)	100
Term	29 - 53	41 (6)	108
1 day pp	18 - 58	38 (10)	100
6 week pp	28 - 68	48 (10)	126

Non-pregnant control reference range - data from OC users excluded

I.F.16. TRANSFERRIN

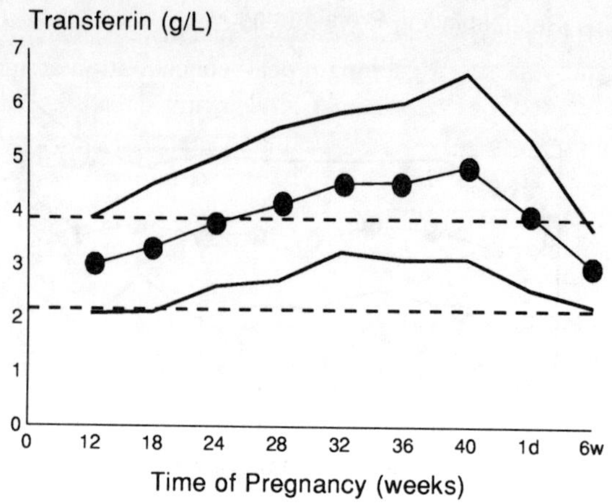

Figure I.F.16. Gestational changes in transferrin concentration.

Table I.F.16. Transferrin (g/L)			
Time of Gestation	Reference Range	Mean (sd)	Percentage of Control mean
Non Pregnant Control	2.18 - 3.89 #	2.85 (0.40)	
12 weeks	2.09 - 3.89	2.99 (0.45)	105
18 weeks	2.13 - 4.53	3.33 (0.60)	117
24 weeks	2.61 - 5.01	3.81 (0.60)	134
28 weeks	2.72 - 5.56	4.14 (0.71)	145
32 weeks	3.25 - 5.85	4.55 (0.65)	160
36 weeks	3.11 - 6.03	4.57 (0.73)	160
Term	3.13 - 6.57	4.85 (0.86)	170
1 day pp	2.57 - 5.37	3.97 (0.70)	139
6 week pp	2.25 - 3.69	2.97 (0.37)	104

Non-pregnant control reference range - data from OC users excluded

I.G. α-Fetoprotein

α-FETOPROTEIN IN PREGNANCY

α-Fetoprotein is the principal plasma protein in early fetal life, synthesized in fetal liver and yolk sac. Reaching a peak concentration at around 12 to 14 weeks gestation, its concentration in fetal serum decreases with increasing gestation.[1] α-Fetoprotein is normally found in amniotic fluid at a concentration 150 times lower than fetal serum. A peak α-fetoprotein concentration is reached in amniotic fluid around 14 weeks and thereafter amniotic fluid α-fetoprotein levels also decrease as pregnancy progresses (Figure and Table I.G.1.).[1]

The normal concentration of α-fetoprotein in maternal serum during pregnancy is 100 to 250 times lower than in amniotic fluid. Despite the drop in concentration in fetal blood, as the fetus grows and blood volume increases, the total fetal α-fetoprotein production increases. This is reflected in increasing α-fetoprotein levels in maternal serum with advancing pregnancy (Figure and Table I.G.1.).

In pathological conditions where α-fetoprotein exudes through a skin defect, very high levels of α-fetoprotein may be found in amniotic fluid and maternal serum.[1-3] These conditions include anencephaly, open spina bifida, gastroschisis, omphalocele, or any other conditions with a break in the skin. Second trimester screening of maternal serum α-fetoprotein is used to identify pregnancies at high risk for these conditions, and multiple pregnancy.[1,4] The finding of high amniotic fluid α-fetoprotein and elevated amniotic fluid acetylcholinesterase can confirm the diagnosis of open neural tube defects.[5]

With the discovery that low maternal serum α-fetoprotein levels were seen in pregnancies where the fetus was chromosomally abnormal,[6,7] maternal serum α-fetoprotein screening for Down syndrome was introduced. The finding that low levels of unconjugated estriol[8-10] and elevated levels of human chorionic gonadotrophin[10-12] also occur more frequently in maternal sera in Down syndrome pregnancies can greatly increase the effectiveness of antenatal screening for Down syndrome.[13]

FACTORS AFFECTING α-FETOPROTEIN CONCENTRATION IN MATERNAL SERUM

When maternal serum α-fetoprotein is used to screen for either neural tube defects or chromosomal abnormalities in pregnancy, there are several factors to consider in interpreting the significance of a given serum α-fetoprotein concentration. Some of the causes of altered serum α-fetoprotein levels are shown in Table I.G.2.

RACE

Blacks have higher serum and probably amniotic fluid concentrations of α-fetoprotein than Caucasians but the prevalence of open spina bifida is lower and the prevalence of Down syndrome similar to Caucasians. Concentrations of α-fetoprotein are similar in Asian and Oriental populations to Caucasians after weight correction factors are applied.

MATERNAL WEIGHT

Circulating plasma volume is greater in larger individuals but fetal derived α-fetoprotein is similar for normal fetuses of similar gestation. Therefore, women with larger plasma volume will have relatively lower concentrations of α-fetoprotein. In calculation of the multiple of the median (MOM), weight correction is probably accurate between a maternal weight of 90 and 280 lbs. No correction for maternal weight is made for amniotic fluid α-fetoprotein since amniotic fluid volume is independent of maternal weight. Used for calculating serum MOM, weight correction increases the detection rate of neural tube defects and decreases the false positive rate.[14,15]

DIABETES

Insulin dependent diabetics have low median values for α-fetoprotein but a higher prevalence for neural tube defects.

SCREENING FOR NEURAL TUBE DEFECTS AND DOWN SYNDROME

Maternal serum α-fetoprotein screening is currently performed for two indications, i.e., screening for open neural tube defects (open spina bifida, OSB) and abdominal wall defects (high MSAFP) and screening for Down syndrome (low MSAFP).

MATERNAL SERUM α-FETOPROTEIN SCREENING FOR NEURAL TUBE DEFECTS

USE OF MULTIPLE OF MEDIAN (MOM) IN CALCULATION OF INDIVIDUAL RISK

The relationship of α-fetoprotein concentrations relative to median serum α-fetoprotein concentrations at each week of gestation to the general population risk for open spina bifida was derived from singleton pregnancies in non-diabetic Caucasian women of normal weight. A different risk calculation is required for populations with a different prevalence rate to the general population. In British Columbia the prevalence rate is approximately 1:700.

The population prevalence is higher in diabetic women and lower in the black population. The prevalence factor to calculate risk of open spina bifida in the black population is 0.50. To calculate risk for a black pregnancy, the general population prevalence (GPP) is multiplied by 0.50. The prevalence factor to calculate risk of OSB in the diabetic population is 3.50. To calculate risk for a diabetic pregnancy, the GPP is multiplied by 3.50. To calculate risk for a diabetic pregnancy in a black woman, the GPP is multiplied by 1.75 (3.50 × 0.5). There are interpretive programs available which can facilitate reporting.

Since absolute values for α-fetoprotein in serum or amniotic fluid may vary between laboratories using different test methods, risk assessment is based on multiples of the median value in a specific population tested by a specific laboratory. Median values are derived from the values found in normal pregnancies at a given stage of gestation.

DIAGNOSTIC EFFICACY OF MATERNAL SERUM α-FETOPROTEIN SCREENING

The sensitivity (true positives detected in all positive cases) and specificity (true negative screens for all negative cases) can be affected by selection of the median cutoff for a positive screen. Thus, by using a cutoff of 2.0 MOM between 80 to 85% of neural tube defects may be detected but the false positive rate may be 5 to 7%. By increasing the cutoff point to 2.5 MOM, fewer neural tube defects may be detected (70 to 75%), but there will be fewer false positives, around 3%.

Several groups have examined outcome in pregnancies where a maternal serum α-fetoprotein for a given gestational age in a singleton pregnancy was found to be high despite absence of a neural tube defect, abdominal wall defect, or other cause as shown in Table 1.G.2.[16,17] An increased incidence of low birth weight, preterm delivery, intrauterine growth retardation, or late pregnancy complication was found, particularly if repeat testing was also high.[16,17] This suggests that an α-fetoprotein level over 2.5 MOM, after gestational age has been confirmed, should alert the physician to a possible high-risk pregnancy, even in the absence of obvious fetal abnormality.

AMNIOTIC FLUID TESTING FOR NEURAL TUBE DEFECTS

Amniotic fluid α-fetoprotein may be measured either initially in those patients who proceed directly to amniocentesis for advanced maternal age (≥35 years) or other appropriate indication or for follow-up of an elevated maternal serum α-fetoprotein. Some of the conditions that may alter amniotic fluid α-fetoprotein levels are shown in Table I.G.3.

If amniotic fluid α-fetoprotein is high (≥2.5 MOM), amniotic fluid acetylcholinesterase may be measured for confirmation of neural tube defect.[5] Acetylcholinesterase, an enzyme secreted by cells of neural origin, i.e., ganglion cells, neurons, and motor nerves), may leak into amniotic fluid from an open spinal defect. This may be used as a confirmatory test, in particular when detailed ultrasound examination does not reveal an obvious defect.

MATERNAL SERUM α-FETOPROTEIN SCREENING FOR DOWN SYNDROME

It is well recognized that the risk of Down syndrome increases with increasing maternal age. At 35 years, the risk of Down syndrome (1:250) was judged to equal or exceed the risk of amniocentesis. Amniocentesis has thus been offered to women who will be ≥35 years at the time of delivery but less than half of eligible women undergo this procedure. Therefore, although by screening this at-risk group, about 5% of pregnancies, 30% of Down syndrome potentially could be detected, less than 15% of affected pregnancies are actually found.[12] About 75% of Down syndrome pregnancies occur in the under 35-year group who are not eligible on the basis of age for amniocentesis.

α-FETOPROTEIN

The concentration of α-fetoprotein in maternal serum was found to be lower than the population medians in pregnancies with Down syndrome.[6,7] Cord serum α-fetoprotein from fetuses with Down syndrome was also low, possibly reflecting decreased fetal hepatic synthesis of α-fetoprotein.[7]

Screening of maternal serum α-fetoprotein for low values was introduced to identify at-risk pregnancies who would then be eligible for amniocentesis.[6,7]

The *a priori* risk for Down syndrome differs by age. A single cutoff of MOM to indicate risk would not differentiate the lower risk group. Since the risk of Down syndrome is so much lower in younger women, an adjusted cutoff is used to determine age-equivalent risks. By adjusting the cutoff of MOM by age, one can detect women with risk equal or greater than that of a 35-year-old (i.e., age-equivalent risk) and therefore justify amniocentesis. An MOM of 0.34 at 21 years implies the same degree of risk as an MOM of 0.63 at 33 years.

If 5% of pregnant women are screened on the basis of age alone, potentially 30% of Down syndrome pregnancies could be detected.[12] Adding α-fetoprotein or other markers such as human chorionic gonadotrophin or unconjugated estriol could increase the detection rate and decrease the false-positive rate.[12]

UNCONJUGATED ESTRIOL

Unconjugated estriol concentrations in maternal sera normally increase in a log-linear fashion by about 20% per week between 15 to 20 weeks gestation.[9] Values may be 15% lower in smokers than non-smokers.[9] Pregnancies with fetal Down syndrome have significantly lower maternal serum unconjugated estriol in the second trimester.[8-10] Using age, maternal serum α-fetoprotein, and maternal serum unconjugated estriol in combination can decrease the false-positive rate to 5% with a detection rate of 45%.

HUMAN CHORIONIC GONADOTROPHIN (HCG)

Human chorionic gonadotrophin in maternal sera comes from the placenta and levels decrease sharply between 10 and 20 weeks of pregnancy.[12,18] In pregnancies with fetal Down syndrome, maternal serum HCG levels are significantly higher than in normal pregnancies.[8,9,14,15] Serum HCG was recently shown to be the best of the three serum markers to distinguish the Down syndrome pregnancy from unaffected pregnancies.[18]

USE OF THESE TESTS IN PREGNANCY: THE "TRIPLE MARKER SCREEN"

While in some European countries, some states in the U.S., and one Canadian province MSAFP has been officially designated as a screening test in pregnancy, this has not yet become routine throughout North America. With the above-mentioned results, it may well become standard practice to perform routine triple marker screening (MSAFP, Estriol, HCG) in all pregnancies.[13,18]

REFERENCES

1. **Knight, G. J.,** Maternal serum α-fetoprotein screening, in *Techniques in Diagnostic Human Biochemical Genetics: A Laboratory Manual,* John Wiley & Sons, New York, 1991, 491.
2. **Brock, D. J. H., Scrimgeour, J. B., and Nelson, M. M.,** Amniotic fluid alpha-fetoprotein measurements in the early prenatal diagnosis of central nervous system disorders, *Clin. Genet.,* 7, 163, 1975.
3. **Kunz, J. and Schmid, J.,** Amniotic alpha-fetoprotein and omphalocoele, *Lancet,* i, 47, 1976.
4. **Adams, M. J., Jr., Windham, G. C., Greenberg, J. F., Clayton-Hopkins, J. A., Reimer, C. B., and Oakley, G. P., Jr.,** Clinical interpretation of maternal serum α-fetoprotein concentrations, *Am. J. Obstet. Gynecol.,* 148, 241, 1984.
5. **Milunsky, A. and Saperstein, V. S.,** Prenatal diagnosis of open neural tube defects using amniotic fluid acetylcholinesterase assay, *Obstet. Gynecol.,* 59, 1, 1982.
6. **Merkatz, I. R., Nitowsky, H. M., Macri, J. N., and Johnson, W. E.,** An association between low maternal serum alpha-fetoprotein (AFP) and fetal chromosomal abnormalities, *Am. J. Obstet. Gynecol.,* 148, 886, 1984.
7. **Cuckle, H. S., Wald, N. J., and Lindenbaum, R. H.,** Maternal serum α-fetoprotein measurement: a screening test for Down syndrome, *Lancet,* i, 926, 1984.

8. **Wald, N. J., Cuckle, H. S., Densem, J. W., Nanchahal, K., Canick, J. A., Haddow, J. E., Knight, G. J., and Palomaki, G. E.,** Maternal serum unconjugated oestriol as an antenatal screening test for Down's syndrome, *Br. J. Obstet. Gynaecol.*, 95, 334, 1988.
9. **Canick, J. A., Knight, G. J., Palomaki, G. E., Haddow, J. E., Cuckle, H. S., and Wald, N. J.,** Low second trimester maternal serum unconjugated oestriol in pregnancies with Down's syndrome, *Br. J. Obstet. Gynaecol.*, 95, 330, 1988.
10. **Chard, T.,** Biochemistry and endocrinology of the Down's syndrome pregnancy, *Ann. N.Y. Acad. Sci.*, 626, 580, 1991.
11. **Bogart, M. H., Pandian, M. R., and Jones, O. W.,** Abnormal maternal serum chorionic gonadotrophin levels in pregnancies with fetal chromosomal abnormalities, *Prenatal. Diagn.*, 7, 623, 1987.
12. **Wald, N. J., Cuckle, H. S., Densem, J. W., Nanchahal, K., Royston, P., Chard, T., Haddow, J. E., Knight, G. J., Palomaki, G. E., and Canick, J. A.,** Maternal serum screening for Down's syndrome in early pregnancy, *Br. Med. J.*, 297, 883, 1988.
13. **Canick, J. A.,** Screening for Down syndrome using maternal serum alpha-fetoprotein, unconjugated estriol, and HCG, *J. Clin. Immunoassay,* 13, 30, 1990.
14. **Johnson, A. M., Palomaki, G. E., and Haddow, J. E.,** The effect of adjusting maternal serum α-fetoprotein levels for maternal weight in pregnancies with fetal open spina bifida, *Am. J. Obstet. Gynecol.*, 163, 9, 1990.
15. **Haddow, J. E., Kloza, G. H., Knight, G. H., and Smith, D. E.,** Relation between maternal weight and serum alpha-fetoprotein concentration during the second trimester, *Clin. Chem.*, 27, 133, 1981.
16. **Hamilton, M. P. R., Abdalla, H. I., and Whitfield, C. R.,** Significance of raised maternal serum α-fetoprotein in singleton pregnancies with normally formed fetuses, *Obstet. Gynecol.*, 65, 465.
17. **Robinson, L., Grau, P., and Crandall, B. F.,** Pregnancy outcomes after increasing maternal serum alpha-fetoprotein levels, *Obstet. Gynecol.*, 74, 17, 1989.
18. **MacDonald, M. L., Wagner, R. M., and Slotnick, R. N.,** Sensitivity and specificity of screening for Down syndrome with alpha-fetoprotein, hCG, unconjugated estriol, and maternal age, *Obstet. Gynecol.*, 77, 63, 1991.

I.G.1. MEDIAN VALUES FOR α-FETOPROTEIN IN MATERNAL SERUM AND AMNIOTIC FLUID

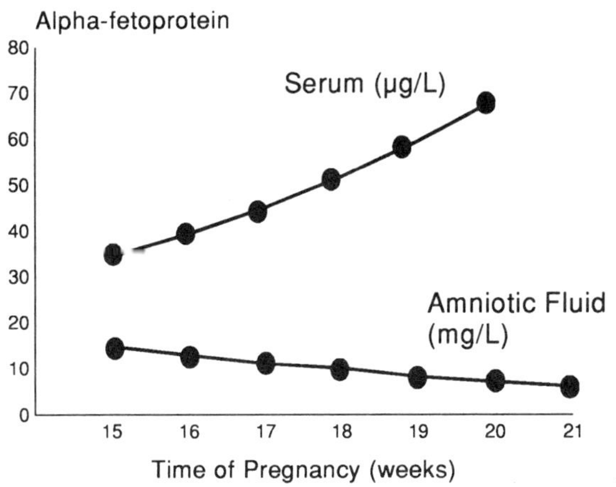

Fig. I.G.1. Changing medians of maternal serum α-fetoprotein and amniotic fluid α-fetoprotein between 15 and 21 weeks of pregnancy.

Table I.G.1. Median Values for α-fetoprotein in maternal serum and amniotic fluid at the British Columbia Children's Hospital laboratory

Week of Pregnancy	Serum μg/L	Amniotic Fluid mg/L
15	34.3	14.7
16	39.3	12.7
17	45.1	10.9
18	51.8	9.9
19	59.4	8.1
20	68.2	7.0
21		6.0

I.G.2. CONDITIONS WHICH ALTER CONCENTRATION OR AFFECT INTERPRETATION OF SERUM α-FETOPROTEIN (MSAFP) LEVELS

Table I.G.2. Conditions which alter concentration or affect interpretation of serum α-fetoprotein (MSAFP) levels

Elevated MSAFP	Low MSAFP
Neural tube defects: open spina bifida, anencephaly, encephalocele	Down syndrome
Underestimated gestational age	Overestimated gestational age
Very low maternal weight	Very high maternal weight
Levels are 10 % higher in black women for given gestation	Levels are lower in women with insulin dependent diabetes mellitus
Multiple pregnancy	
Fetal death Spontaneous abortion, particularly in twin pregnancy	Hydatidiform mole or pseudopregnancy
Other anomalies: 1. gastroschisis or omphalocele 2. congenital nephrosis 3. duodenal or esophageal atresia 4. oligohydramnios 5. hydrocephalus	Other chromosomal anomalies: Trisomy 13 and 18 have been variably described as high or low probably depending on whether associated neural tube defect or anterior wall defect is present or not

I.G.3. CONDITIONS WHICH AFFECT CONCENTRATION OR ALTER INTERPRETATION OF AMNIOTIC FLUID α-FETOPROTEIN (AFAFP) LEVELS

Table I.G.3. Conditions which affect concentration or alter interpretation of amniotic fluid α-fetoprotein (AFAFP) levels

Elevated AFAFP	Low AFAFP
Neural tube defects: open spina bifida, anencephaly, encephalocele	Polyhydramnios
Overestimated gestational age	Underestimated gestational age
Contamination of amniotic fluid specimen with fetal blood	
Multiple pregnancy	
Fetal death Spontaneous abortion	Hydatidiform mole or pseudopregnancy
Other anomalies: 1. gastroschisis or omphalocele 2. congenital nephrosis 3. duodenal or esophageal atresia 4. oligohydramnios 5. hydrocephalus	

I.H. Lipids and Lipid-Soluble Vitamins

EFFECT OF PREGNANCY ON LIPID METABOLISM

Changes in lipid metabolism during pregnancy promote accumulation of maternal fat stores (about 4 kg or 35000 kcal) in early and mid-pregnancy and fat mobilization with enhanced fetal nutrient transfer in the third trimester.[1] In addition, they supply the placenta with cholesterol needed for steroid hormone synthesis and fatty acids for placental oxidation and biomembrane formation.[2-4]

Estrogen contributes to differences in lipoprotein concentrations between males and premenopausal females, and these differences are exaggerated in women using estrogen-containing oral contraceptive preparations and in pregnancy. Estrogens stimulate synthesis of apolipoprotein A-I and VLDL apolipoprotein B with corresponding increases in circulating HDL_2, VLDL, and triglycerides.[5] An estrogen-mediated decrease in heparin-releasable hepatic lipase activity may contribute to this pattern by decreasing turnover of these proteins. Estrogen decreases LDL concentration by increasing hepatic LDL (apo B/E) receptors which clear VLDL remnants and LDL.[2,6]

Synthetic progestins counteract estrogen effects on lipid metabolism to a variable degree, depending on their androgenicity. In most users of combined oral contraceptive preparations the estrogen effect on triglycerides predominates, but changes in HDL_2 and LDL vary with the preparation and may be minimal in preparations with more androgenic progestins. Natural C-21 progestins have less antiestrogenic activity.[2,6]

The variation in serum lipid concentrations during pregnancy has been documented in a number of studies.[7-12] Early changes are consistent with the effects of estrogen, but the subsequent pattern reflects more complex hormonal interactions.

CHOLESTEROL

Cholesterol turnover in non-pregnant women is 1.0 to 1.2 g/day, and this is probably similar in pregnancy. Tissue requirements for cholesterol are altered in pregnancy because the placenta uses large amounts for steroid hormone synthesis.[2] The pattern of change in total cholesterol concentration reflects the changes in cholesterol in the various lipoprotein fractions.

HDL CHOLESTEROL

HDL cholesterol (mainly HDL_2) increases by 12 weeks of pregnancy as expected in response to estrogen, and remains increased thereafter, with maximum values at about 20 weeks.[7,9-11] In our population there was a triphasic pattern: HDL cholesterol concentrations increased to a maximum at 24 weeks of pregnancy, decreased somewhat until 32 weeks, then increased again until term. Proposed explanations for decreasing HDL cholesterol in the second half of pregnancy include increased transfer of cholesterol from HDL to VLDL, a diabetes-like effect on HDL cholesterol secondary to insulin resistance, or increased placental uptake of cholesterol from HDL.[2] The triphasic pattern has also been noted by Desoye et al.[9] They found that HDL_3 accounted for higher HDL cholesterol levels in late pregnancy, and suggested that this reflected accumulation of surface remnants secondary to enhanced lipolysis.[9]

After delivery, despite loss of the estrogen effect, HDL cholesterol remains in the pregnant range for 6 or more weeks in lactating women (our data and others).[11,13] As in late pregnancy, higher HDL cholesterol concentrations during lactation may reflect increased VLDL catabolism (for milk lipid synthesis) with production of surface remnants (HDL_3).[13]

LDL CHOLESTEROL

Total and LDL cholesterol decrease in the early weeks of pregnancy:[8-10] a response consistent with estrogen effects and changes in plasma volume. By 12 weeks there is a balance between decreased LDL and increased HDL cholesterol, and total cholesterol concentrations return to pre-pregnant values. LDL (and total) cholesterol then increase over the second and third trimesters, in parallel with concentrations of cortisol,[12] estradiol,[9,12] progesterone, and human placental lactogen (HPL).[9] The trend in LDL cholesterol is thought to be due to conversion of increased circulating VLDL to LDL (estrogen mediated), which is of greater magnitude than the opposing placental uptake of LDL and HDL cholesterol for steroid hormone synthesis.[2] The effects of individual hormones on LDL cholesterol concentrations have not been determined, but the striking parallel changes in cortisol and cholesterol concentrations, particularly during the stress of labor and delivery,[12] suggest that the role of cortisol in lipid metabolism in pregnancy should be further studied.

After delivery cholesterol decreases slightly, and slowly returns to pre-pregnant values over the next months.[7,8,13] Diagnostic lipid studies should be postponed until 6 months postpartum.[8]

TRIGLYCERIDES

Triglyceride concentrations decrease in the first 8 weeks of pregnancy, possibly reflecting plasma volume expansion and insulin-mediated lipid utilization. By 12 weeks VLDL production has increased under the influence of estrogen and circulating VLDL and triglycerides continue to increase consistently until delivery. In the second half of gestation clearance of circulating VLDL is also altered, due to progressively decreasing activity of lipoprotein lipase in adipose tissue and liver.[2] In contrast, lipoprotein lipase activity increases in placenta and mammary gland, allowing these tissues to use circulating maternal triglycerides.[14] The influence of pregnancy on hepatic lipoprotein lipase represents an interesting physiological adaptation: hepatic enzyme activity is low in the fed state, minimizing hepatic fat utilization, but activity increases above non-pregnant values with fasting, increasing free fatty acid and ketone production to nourish the fetus when the supply of glucose is low.[14,15]

In our population serum triglyceride concentrations at 12 weeks pregnancy were similar to the range in oral contraceptive users, with mean values about 140% of the control mean. The subsequent pattern of change (with approximately a threefold increase by term) is similar in most studies,[7-11] although our results are somewhat higher than those reported for samples collected after longer periods of fasting.[8-11]

Serum triglyceride reference ranges must be interpreted with caution. The serum concentration distribution is skewed, but our reference limits for pregnant women are based on mean and standard deviations because the number of pregnant women studied was insufficient for valid non-parametric reference ranges. The lower reference limit is underestimated. Using log-transformed data to calculate 2 SD reference limits yielded upper and lower reference limits 0.2 to 0.5 mmol/L higher than those illustrated. The effect of calculation method on triglyceride reference range is illustrated by the ranges for non-oral contraceptive users (compare Tables III.B.3. and III.B.4.).

Triglyceride concentrations in the immediate postpartum period decrease by about 50%, followed by a more gradual return to pre-pregnant values over weeks to months.[7,8] The early postpartum decrease may be due to acute effects of prostaglandins on the estrogen/progesterone ratio.[16] Lactation may influence the later postpartum changes, but this is controversial.[8,13]

As for cholesterol, postponement of diagnostic triglyceride measurements until 6 months postpartum is recommended.[8]

LIPOPROTEINS

APOLIPOPROTEINS A-I, A-II, AND B

Hepatic synthesis of apolipoproteins A-I and B increases in pregnancy under the influence of estrogen. Apo-A-I peaks and plateaus at 20 to 25 weeks gestation with values about 25 to 35% higher than for non-pregnant women.[2,9] Apo A-II increases only slightly (12%)[9] or not at all[2] during gestation. Apo B shows the largest increase during gestation, reaching a maximum approximately 75% above non-pregnant values by 32 weeks gestation.[9]

Apoprotein concentrations decrease by 6 weeks postpartum in non-lactating women, but apo A-I and apo A-II concentrations are higher in lactating women, consistent with the sustained increase in HDL cholesterol and HDL during lactation.[13]

LIPOPROTEIN(A)

The pattern of lipoprotein(a) changes in pregnancy is quite different from other lipoproteins. It reaches maximum concentrations approximately three times non-pregnant values by 19 weeks pregnancy and decreases thereafter to baseline levels by term. The pattern has been likened to that of HCG, delayed by 11 weeks, but no definite relationship has been established.[17]

FREE FATTY ACIDS AND GLYCEROL

Plasma free fatty acids and glycerol decrease in early pregnancy, then increase above pre-pregnancy levels in the third trimester.[14,16] The increases in late gestation are consistent with increased lipid mobilization from adipose tissue. Fatty acid mobilization during fasting (discussed above) provides an alternate source of energy for maternal metabolic needs, allowing more glucose as well as fatty acids and ketones to be transferred to the fetus. In the fed state, placental lipoprotein lipase liberates fatty acids that are rapidly transferred to the fetus to supply its needs during this stage of accelerated growth and fat deposition. Excess fatty acids are used by the placenta for placental triglyceride synthesis, cholesterol esterification, membrane biosynthesis, or energy production (oxidation). Placental fatty acid utilization may protect the fetus from excessive fatty acid loads, as in maternal fasting. Partial oxidation of long-chain fatty acids to shorter-chain-length fatty acids by the placenta may also facilitate fetal utilization since the fetus lacks carnitine.[2-4]

PHOSPHOLIPIDS

By 8 weeks pregnancy, phospholipid concentrations are 10 to 15% lower than non-pregnant values, but the steady increase thereafter parallels changes in total cholesterol and lipoprotein concentrations.[9,10] The composition of phospholipids also changes in late pregnancy: lysolecithin decreases and sphyngomyelin, lecithin, and phosphatidyl ethanolamine increase. The significance of these changes is unknown.[16] Phospholipids do not cross the placenta but may be hydrolyzed and reesterified for transfer to the fetal circulation.[2]

DISORDERS OF LIPID METABOLISM IN PREGNANCY

Changes in lipid metabolism in pregnancy make it very difficult to diagnose lipid disorders, especially during the second and third trimesters. Lipid concentrations outside appropriate gestational reference limits suggest an underlying abnormality,[18] but definitive investigation should be postponed until 6 months postpartum.[8] There are a few reports of the effects of pregnancy on lipids in patients with various disorders of lipid metabolism.

Type I hyperlipidemia refers to hyperchylomicronemia due to deficiency or inhibition of lipoprotein lipase or apolipoprotein C-II deficiency. Several patients with lipoprotein lipase deficiency have been followed through pregnancy and lactation.[19,20] During pregnancy plasma triglycerides increased to three to four times the high pre-pregnant levels, despite dietary fat restriction. One patient developed pancreatitis. Infants were normal, with normal cord blood triglycerides, but milk lipid composition was abnormal, reflecting defective mammary gland lipoprotein lipase. Milk lipids are formed by *de novo* fatty acid synthesis rather than from circulating maternal triglycerides. The milk contains less total triglyceride and fatty acids than normal milk, relatively more C_{12} and C_{14} fatty acids, and very low linoleic acid concentrations, and is felt to be nutritionally inadequate.[19,20]

Patients with familial hypercholesterolemia have also been followed through pregnancy.[21-23] Cholesterol typically increases by about 50% in homozygotes, the same relative increase as in normolipidemic pregnant women.[21,22] This increase responds to dietary modification.[22] Atypical cholesterol changes have been reported in one heterozygote for familial hypercholesterolemia.[23] Her serum cholesterol spontaneously decreased during pregnancy to normal levels, and xanthomas resolved. The dramatic increase in LDL degradation was presumably an effect of estrogen.[23]

Potential lipid changes in pregnancy in two other types of hyperlipoproteinemia can be deduced from studies of estrogen administration. Type III hyperlipoproteinemia improves with estrogen therapy because receptor-mediated clearance of intermediate density lipoproteins is increased.[6] In contrast, abnormal VLDL synthesis in type V hyperlipoproteinemia is made worse by estrogen therapy.[6] Similar changes are expected during pregnancy.

Secondary hyperlipidemia has been described in some pregnant diabetics. Lipids in patients with IDDM parallel changes in non-diabetic pregnancies, but pregnant women with NIDDM or GDM may develop exaggerated hypertriglyceridemia that persists for many weeks postpartum.[24]

In pregnant hypolipidemic patients the concern is not for maternal lipid concentrations, but rather for the viability of the fetus. The placenta usually uses maternal LDL cholesterol for steroid hormone synthesis,[4] but this is lacking in patients with abetalipoproteinemic phenotypes. Reported estradiol and progesterone concentrations were approximately half normal values throughout gestation in a patient with homozygous hypobetalipoproteinemia, and estriol production failed to increase in the third trimester. Despite the low steroid hormone concentrations, an appropriately grown infant was born at term (42.5 weeks).[25] Steroid hormone concentrations were normal in one patient with abetalipoproteinemia.[26] These reports suggest that the placenta can synthesize sufficient steroid hormones to support a normal pregnancy using cholesterol from other sources such as *de novo* synthesis or maternal HDL.[25] The nutritional adequacy of milk is a concern, however, since the mammary gland appears to have to rely on *de novo* lipid synthesis, as in type I hyperlipoproteinemia (above), and the milk, at least in homozygous hypobetalipoproteinemia, is deficient in essential fatty acids.[27]

EFFECT OF PREGNANCY ON LIPID-SOLUBLE VITAMINS

VITAMIN A

Vitamin A is obtained from plant carotenoids or retinyl esters in animal tissues. It is required for vision, promotion of cell growth and differentiation, maintenance of epithelial tissues and immune response, and reproductive function. Vitamin A circulates bound to retinol-binding protein (RBP) and prealbumin in plasma. Retinol concentrations are controlled by the availability of RBP, which in turn depends on protein, zinc, and vitamin A nutritional status, as well as effects of hormones such as estrogen (see I.F. Retinol Binding Protein). Excess vitamin A is esterified and stored in the liver. In toxicity, liver stores are overloaded and retinyl esters circulate. Serum retinol concentration is not a good marker for toxicity, due to its dependence on RBP, but concentrations of retinyl esters >10% of serum vitamin A suggest hypervitaminosis A.[28,29]

VITAMIN A IN PREGNANCY

During pregnancy, retinol concentrations remain within the reference range for non-pregnant, non-oral contraceptive users, with a slight tendency for mean values to decrease from 32 weeks to term. Other reports also suggest a decreasing trend later in pregnancy,[30-33] but reported retinol concentrations vary considerably depending on the analytical method used.[34-38] Presumably the estrogen effect that increases serum RBP and retinol in oral contraceptive users is counterbalanced by changes in plasma volume, increased urinary RBP excretion,[39] or other factors, during pregnancy. Retinol concentrations decrease relative to RBP, and the mean molar ratio of retinol to RBP is approximately 0.61 at term compared to 0.77 at 6 weeks postpartum and in non-pregnant controls. Decreased molar saturation of RBP in pregnant women has been reported previously.[40] At 6 weeks postpartum retinol and RBP concentrations are higher than non-pregnant values. The explanation for an increase during lactation is not known.[36,41]

Vitamin A is transferred to the fetus either directly, complexed to maternal RBP, or after placental uptake, transient esterification, and subsequent transfer to RBP in the fetal circulation. Fetal concentrations are lower than maternal due to lower RBP. In the third trimester the fetus accumulates hepatic retinol stores.[42] Fetal stores correlate with maternal vitamin A concentrations.[31] Premature infants and infants of vitamin A-deficient mothers are at risk for vitamin A deficiency. Bronchopulmonary dysplasia in premature infants could be related to vitamin A deficiency, and vitamin A supplementation is being assessed as a preventative measure.[43]

Dietary intake of vitamin A in North America is adequate to support the relatively low fetal requirements, and supplementation is not recommended.[42]

Maternal vitamin A deficiency in other parts of the world is associated with preterm birth, intrauterine growth retardation, and decreased birth weight.[31,34,35] Lower vitamin A concentrations have also been noted in women who developed preeclampsia[32] or abruptio placentae.[33] A relationship between vitamin A and fetal growth and maturation is not unexpected, but the role of vitamin A in the obstetric complications requires further study.

Teratogenesis is a serious concern in patients receiving retinoids during pregnancy. It is common with 13-*cis*-retinoic acid (isotretinoin) administration in early pregnancy, and also occurs with high vitamin A intakes (>25,000 IU/day). Similar defects are rare with lower doses of vitamin A.[28,29] Clinical history is the only way to determine teratogenic risk since serum retinol, with its dependence on RBP, does not reflect vitamin A intake or increased stores. Measurement of preconception plasma concentrations to document that this retinoid, which has a long half-life, has been cleared from the circulation, has been recommended for preventing extretinate tetragenicity.

VITAMIN D

Vitamin D, in the active 1,25-dihydroxyvitamin D form, is essential for calcium and phosphate homeostasis. It increases intestinal calcium and phosphate absorption, modifies renal excretion, and stimulates osteoclast and osteoblast activity, calcium mobilization from bone, and synthesis of calcium-binding proteins. Synthesis of 1,25-dihydroxyvitamin D is controlled by parathyroid hormone,[44] serum calcium and phosphate, and 1,25-dihydroxyvitamin D itself, through regulation of renal 1-hydroxylase activity. Other hormones such as estrogens, growth hormone, prolactin, insulin, and calcitonin can also affect this enzyme.[44,45]

Vitamin D is synthesized in the skin by the action of ultraviolet radiation on 7-dehydrocholesterol. Additional vitamin D comes from the diet, as vitamin D_2 (ergocalciferol) from plants, or vitamin D_3 (cholecalciferol) from animal sources. After absorption, vitamin D is hydroxylated in the liver to 25-hydroxyvitamin D, the main circulating form of the vitamin. Plasma concentrations of 25-hydroxyvitamin D provide an index of vitamin D nutritional status. Vitamin D metabolites are extensively bound to vitamin D-binding protein (DBP) in plasma. Variation in DBP affects total plasma concentrations of vitamin D metabolites, since the free forms regulate metabolism.[46]

VITAMIN D IN PREGNANCY

Vitamin D homeostasis is altered during pregnancy to support the increased calcium and phosphorus requirements of the developing fetus and milk production during lactation.[45,47,48] In the third trimester pregnant women have higher concentrations of DBP and both total and free 25-hydroxyvitamin D and 1,25-dihydroxyvitamin D.[46,49] The changes in 1,25-dihydroxyvitamin

D and DBP have been confirmed by others.[50] Similar but lesser changes have been found in oral contraceptive users.[49] Decreased DBP binding of vitamin D metabolites may be an effect of estrogen to enhance calcium absorption in pregnancy.[46,49]

Vitamin D metabolite concentrations at different stages of gestation have seldom been studied. Variable patterns of plasma 25-hydroxyvitamin D have been reported[44,51] and it has been suggested that the changes reflect normal seasonal variation.[52] Plasma 1,25-dihydroxyvitamin D increases as early as 7 weeks gestation[44] and remains 50 to 100% above non-pregnant values until term.[44,51] Amniotic fluid concentration of this metabolite increased significantly at term.[51] Levels of vitamin D metabolites did not correlate significantly with concentrations of pregnancy hormones, PTH, or calcium during gestation, but did correlate with serum phosphate.

In lactating women total and free 1,25-dihydroxyvitamin D decrease to below non-pregnant values by 2 weeks postpartum, then slowly return to the non-pregnant range. Values are slightly higher in women with low calcium intakes. The decrease in vitamin D concentrations implies that vitamin D is less important for increasing calcium absorption in lactation than in pregnancy.[50]

Cord blood concentrations of 25-hydroxyvitamin D are lower than maternal concentrations.[47,48,53] It is not clear whether there is a relationship between maternal and fetal concentrations of 1,25-dihydroxyvitamin D. High doses of 1,25-dihydroxyvitamin D probably cross the placenta, but in normal pregnancy transport may be minimal, and the fetus probably regulates its own vitamin D metabolism based on its calcium needs.[47,48,54] The placenta can convert 25-hydroxyvitamin D to 1,25-dihydroxyvitamin D, but how this process is regulated and its significance for maternal or fetal metabolism are unknown.

Vitamin D deficiency in pregnancy may cause neonatal hypocalcemia and tetany, tooth enamel hypoplasia, congenital rickets, or maternal osteomalacia. Mothers with malabsorption or malnutrition, vegetarians (especially vegans), and women with low milk intakes are at risk for deficiency, particularly in winter and in northern latitudes.[47,55] Routine supplementation with vitamin D is not recommended in pregnancy, but supplements of 5 to 10 μg/day are suggested for women who drink little or no milk.[55] Hypervitaminosis D could occur in pregnant women receiving excessive vitamin D during pregnancy, or in their newborns, but severe toxicity has not been described even in infants of mothers treated with very high doses for vitamin D metabolic disorders or hypoparathyroidism.[54]

VITAMIN E

Metabolism and clinical effects of vitamin E have been reviewed recently.[56] Vitamin E is a generic term for tocopherols and tocotrienols. The most active substance, and usually the most abundant in the circulation, is α-tocopherol. It functions as an anti-oxidant in cell membranes. After absorption, tocopherol is secreted by the liver in VLDL, and remains in the circulation primarily with HDL and LDL, depending on the distribution of total lipids. In pregnant women a greater proportion of circulating tocopherol is associated with VLDL due to the changes in lipoproteins during pregnancy.[57] Cells take up tocopherol released from VLDL and chylomicrons by lipoprotein lipase, or from LDL by the LDL receptor pathway. Uptake by the placenta is not limited, but transport into fetal plasma is restricted by low fetal lipoprotein concentrations. Cord tocopherol concentrations are about 20% of maternal values.[57-59]

VITAMIN E IN PREGNANCY

Vitamin E increases during pregnancy along with lipoproteins. In our population, as in other reports,[38,59,60] tocopherol concentrations increase in parallel with cholesterol to peak at 150 to 200% of non-pregnant values by term. Both the tocopherol:cholesterol ratio, which we use to correct for lipid-related changes[61] in tocopherol concentrations, and the tocopherol:lipid ratio[60] remain fairly constant throughout gestation. Red blood cell tocopherol, a proposed index of tissue stores, decreases after 15 weeks gestation. Whether this reflects depletion of tissue tocopherol or simply redistribution from RBC to lipoproteins in the vascular compartment is uncertain.[60]

Vitamin E deficiency is not common in pregnancy and there is no need for routine vitamin E supplementation.[42] Low maternal vitamin E concentrations have been associated with obstetric or perinatal complications such as abruptio placentae,[33] preeclampsia,[32] and low birth weight.[35,62] Slightly atypical vitamin E concentrations have been reported in pregnancies that spontaneously aborted.[63] From these studies it is impossible to specify a vitamin E concentration that suggests increased risk because vitamin E concentrations in the control groups were also low, presumably due to differences in analytical methods, and plasma lipids were not assessed. The relationship of perinatal complications to maternal vitamin E status requires further study.

Clinical manifestations of vitamin E deficiency have been described only in premature infants (hemolytic anemia) and patients with severe long-standing fat malabsorption (neuromuscular degeneration). Vitamin E deficiency occurs most commonly in patients with pancreatic insufficiency (cystic fibrosis), impaired bile secretion (biliary atresia, severe liver disease), or intestinal fat malabsorption (abetalipoproteinemia). Supplementation of premature infants has been suggested for prevention of a variety of complications including bronchopulmonary dysplasia, retinopathy of prematurity, anemia, and intraventricular hemorrhage, but this is highly controversial. Only the last is still under investigation.[64] Vitamin E toxicity is a concern only in premature infants.

VITAMIN K

Vitamin K is a coenzyme required for activation of a number of proteins by posttranslational γ-carboxylation of glutamic acid residues. The coagulation factors prothrombin and factors VII, IX, and X are most noticeably affected by vitamin K deficiency, but it is also required for activation of proteins C, S, and Z, bone Gla-protein (osteocalcin), bone matrix Gla-protein, and possibly other proteins.[65] Recommended intakes of vitamin K in pregnancy are the same as non-pregnant recommendations.[42]

VITAMIN K IN PREGNANCY

K vitamins have rarely been measured during pregnancy,[66-69] and usually without comparison to non-pregnant controls. Reference ranges are method dependent.[65] In a recent report, concentrations of phylloquinone at term (0.14 to 2.42 μg/L) could be compared to non-pregnant controls (0.17 to 0.68 μg/L).[68] Higher serum vitamin K concentrations are expected in pregnancy because plasma vitamin K is transported with lipoproteins, especially VLDL and chylomicrons. Concentrations are particularly high in hypertriglyceridemia.[68] Plasma lipid concentrations must be considered when interpreting vitamin K concentrations in pregnancy.

Detection of abnormal (des-γ-carboxylated) prothrombin (PIVKA-II) has also been used to assess vitamin K status in pregnancy, but method differences affect results. In a recent comparative study of plasma PIVKA-II and vitamin K as indices of maternal and neonatal vitamin K deficiency, PIVKA-II was felt to be more useful.[69] Prothrombin time is an insensitive test for vitamin K deficiency.

Symptomatic maternal vitamin K deficiency is rare, but neonatal vitamin K deficiency, hemorrhagic disease of the newborn (HDN), is a common problem in unsupplemented newborns, so attention has focused on placental transport of vitamin K. Cord plasma concentrations of vitamin K are typically less than 10% of maternal values at delivery.[66-69] Differences in maternal and cord lipid concentrations account for only some of the discrepancy. Some studies of the response to maternal administration of vitamin K suggest poor or delayed placental vitamin K transport,[4] but results depend on dose and timing of plasma sampling.[70] Marginal neonatal vitamin K reserves are rapidly depleted on usual diets (e.g., breast milk) which contain little vitamin K, and classical HDN develops in the first week of life. HDN presents earlier in infants of mothers deficient in vitamin K, such as those with malnutrition, malabsorption, or those taking drugs that interfere with vitamin K metabolism (anticonvulsants, coumarin anticoagulants, antibiotics). Supplementation of these mothers with vitamin K, at least during the last 2 weeks of pregnancy, may help prevent early HDN,[69,71,72] but this is controversial.[72] Classic HDN is prevented by prophylactic administration of vitamin K to the newborn.[71-73]

REFERENCES

1. **Hytten, F. E. and Thomson, A. M.,** Maternal physiological adjustments, in *Maternal Nutrition and the Course of Pregnancy,* Committee on Maternal Nutrition/Food and Nutrition Board, National Research Council, Eds., National Academy of Sciences, 1970, 41.
2. **Knopp, R. H., Magee, M. S., Bonet, B., and Comex-Coronado, D.,** Lipid Metabolism in pregnancy, in *Principles of Perinatal-Neonatal Metabolism,* Cowett, R. M., Ed., Springer-Verlag, New York, 1991, 177.
3. **Hay, W. W., Jr.,** The placenta. Not just a conduit for maternal fuels, *Diabetes,* 40, 44, 1991.
4. **Coleman, R. A.,** The role of the placenta in lipid metabolism and transport, *Semin. Perinatol.,* 13, 180, 1989.
5. **Schaefer, E. J., Foster, D. M., Zech, L. A., Lindgren, F. T., Brewer, H. B., Jr., and Levy, R. I.,** The effects of estrogen administration on plasma lipoprotein metabolism in premenopausal females, *J. Clin. Endocrinol. Metab.,* 57, 262, 1983.
6. **Crook, D. and Seed, M.,** Endocrine control of plasma lipoprotein metabolism: effects of gonadal steroids, *Bailliere's Clin. Endocrinol. Metab.,* 4, 851, 1990.
7. **Potter, J. M. and Nestel, P. J.,** The hyperlipidemia of pregnancy in normal and complicated pregnancies, *Am. J. Obstet. Gynecol.,* 133, 165, 1979.
8. **Darmady, J. M. and Postle, A. D.,** Lipid metabolism in pregnancy, *Br. J. Obstet. Gynaecol.,* 89, 211, 1982.
9. **Desoye, G., Schweditsch, M. O., Pfeiffer, K. P., Zechner, R., and Kostner, G. M.,** Correlation of hormones with lipid and lipoprotein levels during normal pregnancy and postpartum, *J. Clin. Endocrinol. Metab.,* 64, 704, 1987.
10. **Fahraeus, L., Larsson-Cohn, U., and Wallentin, L.,** Plasma lipoproteins including high density lipoprotein subfractions during normal pregnancy, *Obstet. Gynecol.,* 66, 468, 1985.
11. **Jimenez, D. M., Pocovi, M., Ramon-Cajal, J., Romero, M. A., Martinez, H., and Grande, F.,** Longitudinal study of plasma lipids and lipoprotein cholesterol in normal pregnancy and puerperium, *Gynecol. Obstet. Invest.,* 25, 158, 1988.
12. **Schwertner, H. A., Torres, L., Jackson, W. G., Maldonado, H. A., Whitson, J. D., and Troxler, R. G.,** Cortisol and the hypercholesterolemia of pregnancy and labor, *Atherosclerosis,* 67, 237, 1987.
13. **Knopp, R. H., Walden, C. E., Wahl, P. W., Bergelin, R., Chapman, M., Irvine, S., and Albers, J. J.,** Effect of postpartum lactation on lipoprotein lipids and apoproteins, *J. Clin. Endocrinol. Metab.,* 60, 542, 1985.
14. **Herrera, E., Lasuncion, M. A., Gomez-Coronado, D., Aranda, P., Lopez-Luna, P., and Maier, I.,** Role of lipoprotein lipase activity on lipoprotein metabolism and the fate of circulating triglycerides in pregnancy, *Am. J. Obstet. Gynecol.,* 158, 1575, 1988.
15. **Metzger, B. E., Ravnikar, V., Vileisis, R. A., and Freinkel, N.,** "Accelerated starvation" and the skipped breakfast in late normal pregnancy, *Lancet,* i, 588, 1982.
16. **Winick, M.,** Carbohydrate, fat and protein, in *Nutrition, Pregnancy, and Early Infancy,* Williams & Wilkins, Baltimore, 1989, 46.
17. **Zechner, R., Desoye, G., Schweditsch, M. O., Pfeiffer, K. P., and Kostner, G. M.,** Fluctuations of plasma lipoprotein-A concentrations during pregnancy and post partum, *Metabolism,* 35, 333, 1986.
18. **Frohlich, J. J.,** Laboratory diagnosis of hyperlipidemia, in *Primary Hyperlipoproteinemias,* Steiner, G. and Shafrir, E., Eds., McGraw-Hill, New York, 1991, 143.
19. **Steiner, G., Myher, J. J., and Kuksis, A.,** Milk and plasma lipid composition in a lactating patient with type I hyperlipoproteinemia, *Am. J. Clin. Nutr.,* 41, 121, 1985.

20. **Berger, G. M. B., Spark, A., Baillie, P. M., Huskisson, J., Stockwell, G., and Van de Merwe, E.,** Absence of serum-stimulated lipase activity and altered lipid content in milk from a patient with type I hyperlipoproteinaemia, *Pediatr. Res.*, 17, 835, 1983.
21. **Barss, V., Phillippe, M., Greene, M. F., and Covell, L.,** Pregnancy complicated by homozygous hypercholesterolemia, *Obstet. Gynecol.*, 65, 756, 1985.
22. **McMurry, M. P., Connor, W. E., and Goplerud, C. P.,** The effects of dietary cholesterol upon the hypercholesterolemia of pregnancy, *Metabolism,* 30, 869, 1981.
23. **Mabuchi, H., Sakai, Y., Watanabe, T., Koizumi, J., and Takeda, R.,** Normalization of low-density lipoprotein levels and disappearance of xanthomas during pregnancy in a woman with heterozygous familial hypercholesterolemia, *Metabolism,* 34, 309, 1985.
24. **Hollingsworth, D. R. and Grundy, S. M.,** Pregnancy-associated hypertriglyceridemia in normal and diabetic women, *Diabetes,* 31, 1092, 1982.
25. **Parker, C. R., Illingworth, D. R., Bissonnette, J., and Carr, B. R.,** Endocrine changes during pregnancy in a patient with homozygous familial hypobetalipoproteinemia, *N. Engl. J. Med.*, 314, 557, 1986.
26. **Kayden, H. J. and Traber, M. G.,** Abetalipoproteinemia and homozygous hypobetalipoproteinemia, in *Primary Hyperlipoproteinemias,* Steiner, G. and Shafrir, E., Eds., McGraw-Hill, New York, 1991, 249.
27. **Wang, C. S. and Illingworth, D. R.,** Lipid composition and lipolytic activities in milk from a patient with homozygous familial hypobetalipoproteinemia, *Am. J. Clin. Nutr.*, 45, 730, 1987.
28. **Olson, J. A.,** Vitamin A, in *Handbook of Vitamins,* 2nd ed., Machlin, L. J., Ed., Marcel Dekker, New York, 1991, 1.
29. **Hathcock, J. N., Hattan, D. G., Jenkins, M. Y., McDonald, J. T., Sundaresan, P. R., and Wilkening, V. L.,** Evaluation of vitamin A toxicity, *Am. J. Clin. Nutr.*, 52, 183, 1990.
30. **Howells, D. W., Haste, F., Rosenberg, D., Brown, I. R. F., and Brooke, O. G.,** Investigation of vitamin A nutrition in pregnant British Asians and their infants, *Hum. Nutr. Clin. Nutr.*, 40, 43, 1986.
31. **Shah, R. S., Rajalakshmi, R., Bhatt, R. V., Hazra, M. N., Patel, B. C., Swamy, N. B., and Patel, T. V.,** Liver stores of vitamin A in human fetuses in relation to gestational age, fetal size and maternal nutritional status, *Br. J. Nutr.*, 58, 181, 1987.
32. **Jendryczko, A. and Drozdz, M.,** Plasma retinol, β-carotene and vitamin E levels in relation to the future risk of pre-eclampsia, *Zentralbl. Gynakol.*, 111, 1121, 1989.
33. **Sharma, S. C., Bonnar, J., and Dostalova, L.,** Comparison of blood levels of vitamin A, β-carotene and vitamin E in abruptio placentae with normal pregnancy, *Int. J. Vit. Nutr. Res.*, 56, 3, 1986.
34. **Shah, R. S. and Rajalakshmi, R.,** Vitamin A status of the newborn in relation to gestational age, body weight, and maternal nutritional status, *Am. J. Clin. Nutr.*, 40, 794, 1984.
35. **Navarro, J., Bourgeay Causse, M., Desquilbet, N., Herve, F., and Lallemand, D.,** The vitamin status of low birth weight infants and their mothers, *J. Pediatr. Gastroenterol. Nutr.*, 3, 744, 1984.
36. **Ala-Houhala, M., Koskinen, T., Maki, R., and Rinkari, S.,** Serum vitamin A levels in mothers and their breast-fed term infants with or without supplemental vitamin A, *Acta Paediatr. Scand.*, 77, 198, 1988.
37. **Vobecky, J. S., Vobecky, J., Shapcott, D., Demers, P.-P., Cloutier, D., Blanchard, R., and Fisch, C.,** Biochemical indices of nutritional status in maternal, cord, and early neonatal blood, *Am. J. Clin. Nutr.*, 36, 630, 1982.
38. **Dostalova, L.,** Correlation of the vitamin status between mother and newborn during delivery, *Dev. Pharmacol. Ther.*, 4, 45, 1982.

39. **Beetham, R., Dawnay, A., Menabawy, M., and Silver, A.,** Urinary excretion of albumin and retinol-binding protein during normal pregnancy, *J. Clin. Pathol.*, 41, 1089, 1988.
40. **Jansson, L. and Nilsson, B.,** Serum retinol and retinol-binding protein in mothers and infants at delivery, *Biol. Neonate,* 43, 269, 1983.
41. **Cumming, F. J. and Briggs, M. H.,** Changes in plasma vitamin A in lactating and non-lactating oral contraceptive users, *Br. J. Obstet. Gynaecol.*, 90, 73, 1983.
42. **Subcommittee on Dietary Intake and Nutrient Supplements During Pregnancy,** Vitamins A, E, and K, in *Nutrition During Pregnancy,* Subcommittee on Nutritional Status and Weight Gain During Pregnancy, Subcommittee on Dietary Intake and Nutrient Supplements During Pregnancy, Committee on Nutritional Status During Pregnancy and Lactation, Food and Nutrition Board, Institute of Medicine, National Academy of Sciences, National Academy Press, Washington, DC, 1990, 336.
43. **Zachman, R. D.,** Retinol (vitamin A) and the neonate: special problems of the human premature infant, *Am. J. Clin. Nutr.*, 50, 413, 1989.
44. **Reddy, S. G., Norman, A. W., Willis, D. M., Goltzman, D., Guyda, H., Solomon, S., Philips, D. R., Bishop, J. E., and Mayer, E.,** Regulation of vitamin D metabolism in normal human pregnancy, *J. Clin. Endocrinol. Metab.*, 56, 363, 1983.
45. **Collins, E. D. and Norman, A. W.,** Vitamin D, in *Handbook of Vitamins,* 2nd ed., Machlin, L. J., Ed., Marcel Dekker, New York, 1991, 59.
46. **Bikle, D. D. and Gee, E.,** Measurement of free vitamin D metabolite level: biological and clinical significance, in *Vitamin D. Molecular, Cellular and Clinical Endocrinology,* Norman, A. W., Schaefer, K., Grigoleit, H.-G., and Herrath, D. V., Eds., Walter de Gruyter, Berlin, 1988, 703.
47. **Tsang, R. C., Greer, F., and Steichen, J. J.,** Perinatal metabolism of vitamin D. Transition from fetal to neonatal life, *Clin. Perinatol.*, 8, 287, 1981.
48. **Salle, B. L., Senterre, J., Glorieux, F. H., Delvin, E. E., and Putet, G.,** Vitamin D metabolism in preterm infants, *Biol. Neonate,* 52, 119, 1987.
49. **Bikle, D. D., Gee, E., Halloran, B., and Haddad, J. G.,** Free 1,25-dihydroxyvitamin D levels in serum from normal subjects, pregnant subjects, and subjects with liver disease, *J. Clin. Invest.*, 74, 1966, 1984.
50. **Wilson, S. G., Retallack, R. W., Kent, J. C., Worth, G. K., and Gutteridge, D. H.,** Serum free 1,25-dihydroxyvitamin D and the free 1,25-dihydroxyvitamin D index during a longitudinal study of human pregnancy and lactation, *Clin. Endocrinol.*, 32, 613, 1990.
51. **Issa, S., Mallmann, R., and Burmeister, W.,** Concentrations of vitamin D metabolites in amniotic fluid and serum during pregnancy, in *Vitamin D. Molecular, Cellular and Clinical Endocrinology,* Norman, A. W., Schaefer, K., Grigoleit, H.-G., and Herrath, D. V., Eds., Walter de Gruyter, Berlin, 1988, 917.
52. **van den Berg, H. and Bruinse, H. W.,** Vitamin requirements in normal human pregnancy, *World Rev. Nutr. Diet.*, 57, 95, 1988.
53. **Axelson, M. and Christensen, N. J.,** Vitamin D metabolism in human pregnancy. Concentrations of free and sulphated 25-hydroxyvitamin D_3 in maternal and fetal plasma at term, *J. Steroid Biochem.*, 31, 35, 1988.
54. **Mehls, O., Wolf, H., and Wille, L.,** Vitamin D requirements and vitamin D intoxication in infancy, *Internat. J. Vit. Nutr. Res.*, Suppl. 30, 87, 1989.
55. **Subcommittee on Dietary Intake and Nutrient Supplements During Pregnancy,** Calcium, vitamin D, and magnesium, in *Nutrition During Pregnancy,* Committee on Nutritional Status During Pregnancy and Lactation, Food and Nutrition Board, Institute of Medicine, National Academy of Sciences, National Academy Press, Washington, DC, 1990, 318.

56. **Sokol, R. J.,** Vitamin E and neurologic deficits, *Adv. Pediatr.,* 37, 119, 1990.
57. **Ogihara, T., Miki, M., Kitagawa, M., and Mino, M.,** Distribution of tocopherol among human plasma lipoproteins, *Clin. Chim. Acta,* 174, 299, 1988.
58. **Desai, I. D., Martinez, F. E., Dos Santos, J. E., and Dutra De Oliveria, J. E.,** Transient lipoprotein deficiency at birth: a cause of low levels of vitamin E in the newborn, *Acta Vitaminol. Enzymol.,* 6, 71, 1984.
59. **Haga, P. and Lunde, G.,** Selenium and vitamin E in cord blood from preterm and full term infants, *Acta Paediatr. Scand.,* 67, 735, 1978.
60. **Mino, M. and Nagamatu, M.,** An evaluation of nutritional status of vitamin E in pregnant women with respect to red blood cell tocopherol level, *Int. J. Vit. Nutr. Res.,* 56, 149, 1986.
61. **Farrell, P. M., Levine, S. L., Murphy, D., and Adams, A. J.,** Plasma tocopherol levels and tocopherol-lipid relationships in a normal population of children as compared to healthy adults, *Am. J. Clin. Nutr.,* 31, 1720, 1978.
62. **Shah, R. S., Rajalakshmi, R., Bhatt, R. V., Hazra, M. N., Patel, B. C., Swamy, N. B., and Patel, T. V.,** Vitamin E status of the newborn in relation to gestational age, birth weight and maternal vitamin E status, *Br. J. Nutr.,* 58, 191, 1987.
63. **Vobecky, J. S., Vobecky, J., Shapcott, D., Cloutier, D., Lafond, R., and Blanchard, R.,** Vitamins C and E in spontaneous abortion, *Int. J. Vit. Nutr. Res.,* 46, 291, 1976.
64. **Phelps, D. L.,** The role of vitamin E therapy in high-risk neonates, *Clin. Perinatol.,* 15, 55, 1988.
65. **Suttie, J. W.,** Vitamin K, in *Handbook of Vitamins,* 2nd ed., Machlin, L. J., Ed., Marcel Dekker, New York, 1991, 145.
66. **Hiraike, H., Kimura, M., and Itokawa, Y.,** Distribution of K vitamins (phylloquinone and menaquinones) in human placenta and maternal and umbilical cord plasma, *Am. J. Obstet. Gynecol.,* 158, 564, 1988.
67. **Shearer, M. J.,** Plasma vitamin K_1 in mothers and their newborn babies, *Lancet,* ii, 460, 1982.
68. **Shearer, M. J., McCarthy, P. T., Crampton, O. E., and Mattock, M. B.,** The assessment of human vitamin K status from tissue measurements, in *Current Advances in Vitamin K Research,* Suttie, J. W., Ed., Elsevier, New York, 1988, 437.
69. **Motohara, K., Takagi, S., Endo, F., Kiyota, Y., and Matsuda, I.,** Oral supplementation of vitamin K for pregnant women and effects on levels of plasma vitamin K and PIVKA-II in the neonate, *J. Pediatr. Gastroenterol. Nutr.,* 11, 32, 1990.
70. **Suzuki, S., Maki, M., Shirakawa, K., and Terao, T.,** Metabolism of vitamin K dependent factors in mothers and their newborn infants, *J. Perinat. Med.,* 17, 305, 1989.
71. **Kries, R. V., Shearer, M. J., and Göbel, U.,** Vitamin K in infancy, *Eur. J. Pediatr.,* 147, 106, 1988.
72. **Fetus and Newborn Committee, Canadian Paediatric Society,** The use of vitamin K in the perinatal period, *C.M.A.J.,* 139, 127, 1988.
73. **Hathaway, W. E.,** New insights on vitamin K, *Hematol. Oncol. Clin. North Am.,* 1, 367, 1987.

I.H.1. CHOLESTEROL (TOTAL)

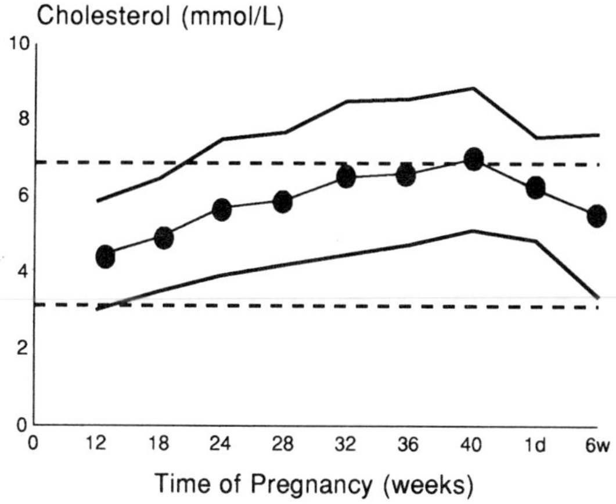

Figure I.H.1. Gestational changes in cholesterol concentration.

Table I.H.1. Cholesterol (mmol/L)

Time of Gestation	Reference Range	Mean (sd)	Percentage of Control mean
Non Pregnant Control	3.13 - 6.87 *	4.48 (0.92)	
12 weeks	3.02 - 5.94	4.48 (0.73)	100
18 weeks	3.51 - 6.47	4.99 (0.74)	111
24 weeks	3.91 - 7.47	5.69 (0.89)	127
28 weeks	4.17 - 7.69	5.93 (0.88)	132
32 weeks	4.45 - 8.49	6.47 (1.01)	144
36 weeks	4.71 - 8.55	6.63 (0.96)	148
Term	5.12 - 8.84	6.98 (0.93)	156
1 day pp	4.83 - 7.55	6.19 (0.68)	138
6 week pp	3.34 - 7.66	5.50 (1.08)	123

* Note upper limit is higher than currently recommended for decreasing risk of atherosclerosis

I.H.2. HDL-CHOLESTEROL

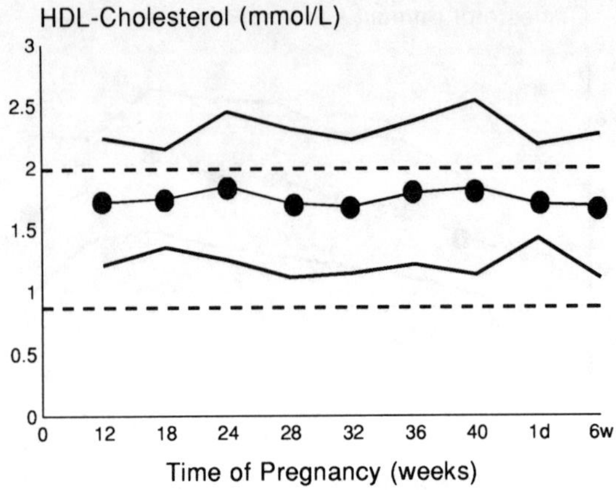

Figure I.H.2. Gestational changes in HDL-cholesterol concentration.

Table I.H.2. HDL-Cholesterol (mmol/L)

Time of Gestation	Reference range	Mean (sd)	Percentage of Control mean
Non Pregnant Control	0.87 - 1.99	1.41 (0.27)	
12 weeks	1.20 - 2.24	1.70 (0.26)	121
18 weeks	1.35 - 2.15	1.75 (0.20)	124
24 weeks	1.25 - 2.45	1.85 (0.30)	131
28 weeks	1.11 - 2.31	1.71 (0.31)	121
32 weeks	1.14 - 2.22	1.68 (0.27)	119
36 weeks	1.21 - 2.37	1.79 (0.29)	127
Term	1.13 - 2.53	1.83 (0.35)	130
1 day pp	1.42 - 2.18	1.70 (0.24)	116
6 week pp	1.10 - 2.26	1.68 (0.29)	119

I.H.3. LDL-CHOLESTEROL

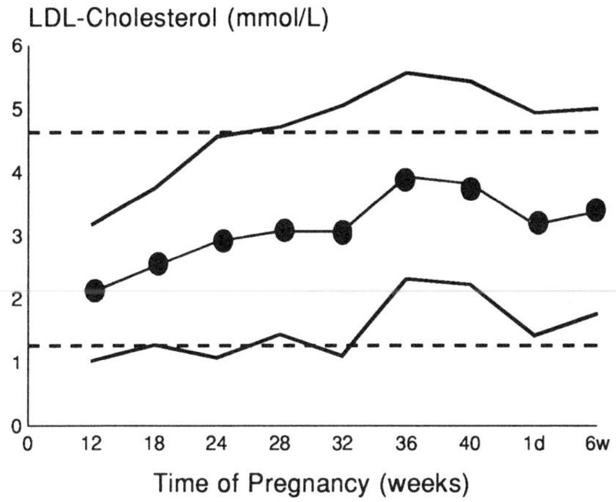

Figure I.H.3. Gestational changes in LDL-cholesterol concentration.

Table I.H.3. LDL Cholesterol (mmol/L). This is a derived value.
LDL-Cholesterol = [total cholesterol - ({ triglyceride/2.2 } + HDL-cholesterol)]

Time of Gestation	Reference Range	Mean (sd)	Percentage of Control mean
Non Pregnant Control	1.27 - 4.64	2.62 (0.83)	
12 weeks	1.02 - 3.18	2.10 (0.54)	80
18 weeks	1.28 - 3.76	2.52 (0.62)	96
24 weeks	1.07 - 4.57	2.91 (0.92)	111
28 weeks	1.44 - 4.72	3.08 (0.82)	118
32 weeks	1.10 - 5.06	3.08 (0.99)	118
36 weeks	2.32 - 5.56	3.94 (0.81)	150
Term	2.23 - 5.43	3.83 (0.80)	146
1 day pp	1.42 - 4.94	3.18 (0.88)	121
6 week pp	1.77 - 5.01	3.39 (0.81)	129

I.H.4. TRIGLYCERIDES

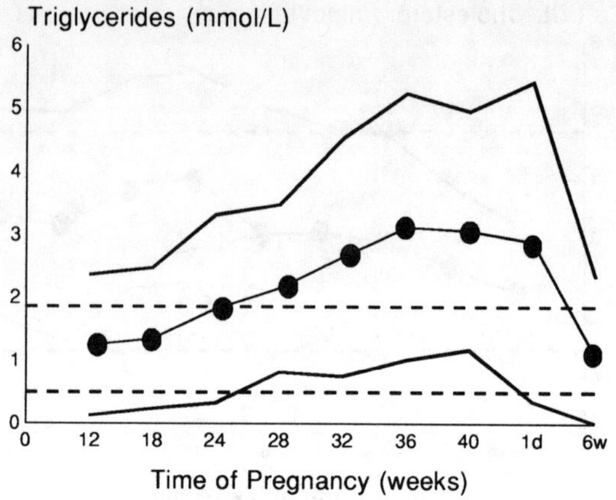

Figure I.H.4. Gestational changes in triglyceride concentration.

Table I.H.4. Triglycerides (mmol/L)			
Time of Gestation	Reference range	Mean (sd)	Percentage of Control mean
Non Pregnant Control	0.49 - 1.86 #	0.88 (0.35)	
12 weeks	0.12 - 2.36	1.24 (0.56)	141
18 weeks	0.23 - 2.47	1.35 (0.56)	153
24 weeks	0.32 - 3.32	1.82 (0.75)	207
28 weeks	0.81 - 3.49	2.15 (0.67)	244
32 weeks	0.74 - 4.54	2.64 (0.95)	300
36 weeks	1.01 - 5.25	3.13 (1.06)	356
Term	1.17 - 4.97	3.07 (0.95)	349
1 day pp	0.35 - 5.43	2.89 (1.27)	328
6 week pp	0 - 2.33	1.07 (0.63)	122

Non-pregnant control reference range - data from OC users excluded

I.H.5. RETINOL

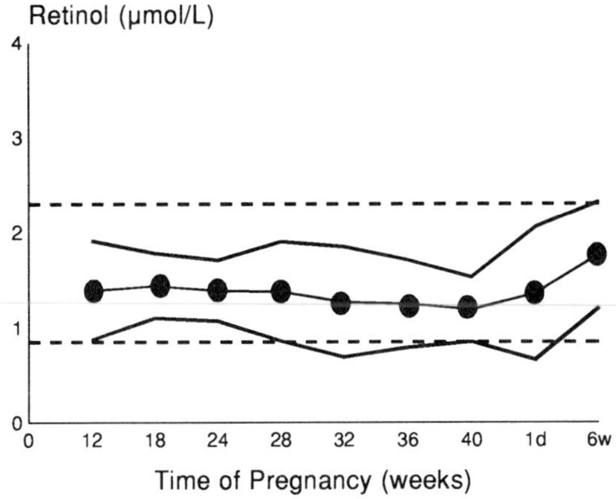

Figure I.H.5. Gestational changes in retinol (vitamin A) concentration.

Table I.H.5. Retinol (µmol/L)			
Time of Gestation	Reference Range	Mean (sd)	Percentage of Control mean
Non Pregnant Control	0.85 - 2.31 #	1.40 (0.34)	
12 weeks	0.87 - 1.91	1.39 (0.26)	99
18 weeks	1.10 - 1.78	1.44 (0.18)	103
24 weeks	1.07 - 1.71	1.39 (0.16)	99
28 weeks	0.86 - 1.90	1.38 (0.27)	99
32 weeks	0.69 - 1.85	1.27 (0.29)	91
36 weeks	0.79 - 1.71	1.25 (0.23)	89
Term	0.85 - 1.53	1.19 (0.18)	85
1 day pp	0.66 - 2.06	1.36 (0.35)	97
6 week pp	1.21 - 2.33	1.77 (0.28)	126

Non-pregnant control reference range - data from OC users excluded

I.H.6. α-TOCOPHEROL

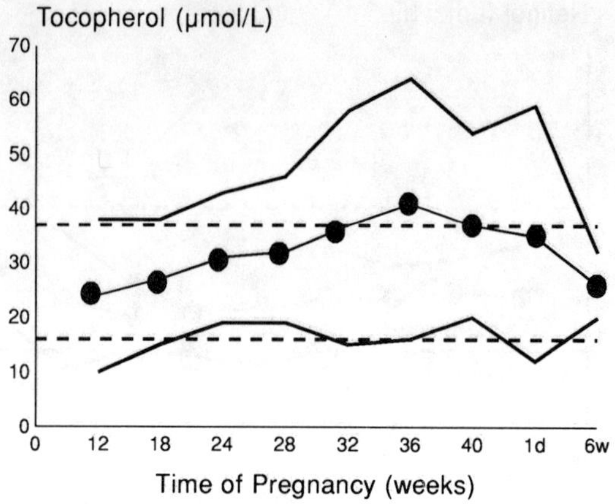

Figure I.H.6. Gestational changes in α-tocopherol concentration.

Table I.H.6. α-Tocopherol (μmol/L)			
Time of Gestation	Reference Range	Mean (sd)	Percentage of Control mean
Non Pregnant Control	16 - 37	24 (5)	
12 weeks	10 - 38	24 (7)	100
18 weeks	15 - 39	27 (6)	113
24 weeks	19 - 43	31 (6)	129
28 weeks	18 - 46	32 (7)	133
32 weeks	15 - 59	37 (11)	154
36 weeks	17 - 65	41 (12)	171
Term	21 - 53	37 (8)	154
1 day pp	11 - 59	35 (12)	146
6 week pp	20 - 32	26 (3)	108

I.H.7. α-TOCOPHEROL: CHOLESTEROL RATIO

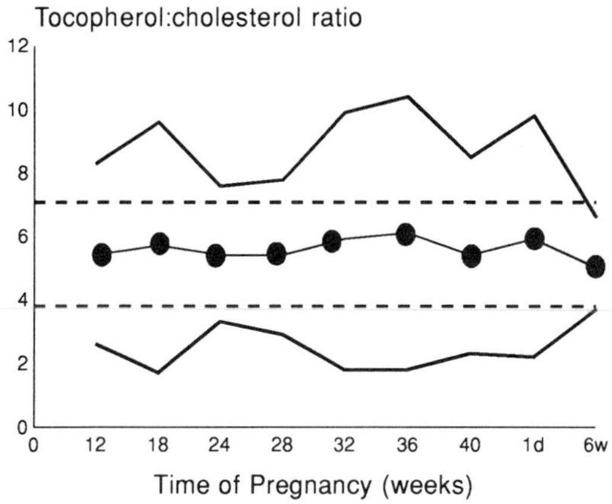

Figure I.H.7. Gestational changes in α-tocopherol:cholesterol ratio.

Table I.H.7. Ratio of α-tocopherol:cholesterol
This is a derived value: Ratio = (α-tocopherol/cholesterol)

Time of Gestation	Reference Range	Mean (sd)	Percentage of Control mean
Non Pregnant Control	3.8 - 7.1	5.2 (1.0)	
12 weeks	2.6 - 8.3	5.4 (1.4)	104
18 weeks	1.7 - 9.6	5.7 (2.0)	110
24 weeks	3.3 - 7.6	5.4 (1.2)	104
28 weeks	2.9 - 7.8	5.4 (1.2)	104
32 weeks	1.8 - 9.9	5.9 (2.0)	113
36 weeks	1.8 - 10.4	6.1 (2.1)	117
Term	2.3 - 8.5	5.4 (1.5)	104
1 day pp	2.2 - 9.8	5.9 (1.9)	113
6 week pp	3.7 - 6.6	5.0 (1.3)	96

I.I. Water-Soluble Vitamins

ASSESSMENT OF VITAMIN NUTRITIONAL STATUS

The water-soluble B group vitamins are a group of compounds which act as specific co-factors for a variety of enzyme reactions in the metabolism of carbohydrates, lipids, and proteins.[1,2] Since these vitamins are not stored to any significant degree and are rapidly excreted in the urine, deficiencies may develop much more rapidly in response to unusual stresses than are seen for the lipid-soluble vitamins.[1]

Single-nutrient deficiencies can arise in specific situations such as unsupplemented parenteral nutrition.[3] More commonly in a malnourished population, multiple vitamin deficiencies occur together.[4] Synergistic metabolic interactions between some vitamins[5] may enhance the effect of multiple deficiencies.

Signs and symptoms of dietary B vitamin deficiency are often rather nonspecific and detection of a vitamin deficiency requires laboratory testing.[1,2,6-8] However, laboratory evaluation of the nutritional status of an individual for many of the water-soluble vitamins is not very satisfactory as there is some controversy over the most useful indices. Blood or plasma concentrations of the vitamins may change very acutely with recent ingestion.[1,6] Alternative methods have been developed to provide a way to assess longer-term nutritional status. For three of these vitamins, thiamin, riboflavin, and pyridoxine, a functional assay of a vitamin-dependent enzyme in the erythrocyte has been used to determine deficiency.[1,2,6] For thiamin status, erythrocyte transketolase activity is measured, for riboflavin, glutathione reductase activity, and, for pyridoxine, erythrocyte asparate or alanine aminotransferase activity.

Ascorbic acid status is commonly assessed by serum ascorbate levels although these vary rapidly with changes in intake.[7] Similar caveats exist for serum folate and B_{12} measurements.[8] Erythrocyte assays for folate indicate longer-term deficiencies but erythrocyte B_{12} assays are less useful.[8] The presence of megaloblastic anemia with hypersegmented neutrophils suggests deficiency, but erythrocyte morphology cannot differentiate between deficiency of folate and vitamin B_{12}.[8] Because of the close interaction between B_{12} and folate, both deficiencies may exist simultaneously and they should be assessed together.

Detailed reviews of methods of assessment of vitamin nutritional status may be found in recent sources.[1,2,7,8]

EFFECT OF PREGNANCY ON VITAMIN STATUS

THIAMIN, RIBOFLAVIN, AND PYRIDOXINE

Thiamin

Thiamin (B_1), as thiamin diphosphate, is an important co-factor in oxidative carboxylation in the pyruvate, α-ketoglutarate, and branched-chain ketoacid dehydrogenase complexes.[9,10] The activity of thiamin as co-factor for the transketolase enzyme in the erythrocyte provides a functional diagnostic test for thiamin deficiency.[1,6,10]

Thiamin deficiency disrupts the oxidative metabolism of glucose and leads to the clinical syndrome due to tissue damage from energy depletion.[9,10] Chronic nutritional thiamin deficiency may present with two main clinical syndromes. Dry beriberi is a form of peripheral neuropathy occurring when thiamin deficiency occurs with low overall calorie intake, whereas wet beriberi, (high output cardiac failure) occurs with severe thiamin deficiency but relatively higher total energy intake.[10]

In developed countries, clinical thiamin deficiency may develop in states of malnutrition associated with alcoholism[10] or with unsupplemented parenteral nutrition.[3] Since thiamin is not stored, deficiency can develop within 2 to 3 weeks of inadequate intake.[1,3] Severe thiamin deficiency with development of Wernicke's encephalopathy has been reported in a woman with severe intractable hyperemesis gravidarum.[11]

Biochemical thiamin deficiency, detected by an increased transketolase stimulation index, has been reported in mothers during pregnancy.[12,13] Although these findings were not correlated with any obvious maternal signs of deficiency, this led to some concern that subclinical maternal thiamin deficiency, demonstrated biochemically, could result in thiamin deficiency in infants who are breast fed by these mothers. Thiamin deficiency with encephalopathy has occurred in an infant on a thiamin-deficient formula.[14]

In the absence of any proof that subclinical thiamin deficiency is a problem in North America, the Canadian guidelines for recommended nutrient intake in pregnancy remains at 0.40 mg thiamin per 1000 kcal daily intake.[15] The American guidelines for recommended dietary allowance suggest that an increased intake of 0.4 mg/day is needed throughout pregnancy.[16]

Riboflavin

Dietary riboflavin is converted to two important coenzymes, flavin mononucleotide (FMN) and flavin adenine dinucleotide (FAD), required for many key metabolic electron-transfer reactions.[17] Because FMN is required for activation of pyridoxine (B_6) and formation of niacin from tryptophan, pure riboflavin deficiency is rare.[4,5,17] Clinical features of pyridoxine, niacin, and riboflavin deficiency include angular stomatitis, glossitis, and dermatitis.[4,17]

Nutritional riboflavin deficiency is common in situations of dietary inadequacy.[17] Biochemical evidence of riboflavin deficiency using tests of vitamin-dependent enzyme activity, has been demonstrated in apparently asymptomatic pregnant women.[18] An increase in intake of 0.3 mg riboflavin per day was suggested in both the Canadian and American guidelines.[15,16]

Pyridoxine

The co-factors pyridoxal-5-phosphate and pyridoxamine-5-phosphate are important in amino acid metabolism. Alanine aminotransferase and aspartate aminotransferase are pyridoxine-dependent enzymes, and this forms the basis for a functional assay for pyridoxine deficiency.[2,6]

Isolated deficiency of pyridoxine on a dietary basis is rare,[5] but pyridoxine metabolism may be altered in hepatic and renal disease.[19] Pyridoxine therapy is effective in treatment of some genetic disorders such as a subgroup of homocystinuria, pyridoxine responsive anemia, and neonatal seizures.[20]

Clinical pyridoxine deficiency may be associated with fatigue, headache, anemia, weight loss, vomiting, hyperirritability, depression, and convulsions.[2] Other associated signs may include dermatitis, glossitis, stomatitis, and cheilosis as described for riboflavin.

Studies have demonstrated elevated erythrocyte aminotransferase stimulation indices in pregnancy suggesting subclinical pyridoxine deficiency.[21,22] A marked decrease in plasma pyridoxal-5′-phosphate concentration has also been described.[23,24] This can be prevented by very large doses of pyridoxine.[23] However, whether the decrease in concentration represents deficiency or is physiological, analogous to that seen for zinc or selenium, has been questioned.[24,25] The Canadian guidelines suggest that no increase in the 0.015 mg B_6 per gram of dietary protein is needed in pregnancy.[15] Since there is an increase in protein needs in pregnancy, this would amount to an additional intake of about 0.6 mg/day.[16]

A recent double-blind, randomized control trial comparing pyridoxine with placebo suggested that ingestion of 25 mg pyridoxine hydrochloride taken 8 hourly for 3 days was effective in reducing the nausea and vomiting of early pregnancy.[26]

Although pyridoxine is relatively non-toxic, peripheral neuropathy has developed after chronic ingestion of pyridoxine supplements.[27,28]

ASSESSMENT OF THIAMIN, PYRIDOXINE, AND RIBOFLAVIN STATUS

Functional assays of vitamin-dependent enzymes in the erythrocyte were used to determine deficiency of thiamin (erythrocyte transketolase activity), riboflavin (glutathione reductase activity), and pyridoxine (erythrocyte aspartate aminotransferase activity).

The principle of these functional assays is as follows: the specific enzyme activity is measured in two reactions, one without additional specific co-factor (basal) and the other with a saturating amount of the co-factor added (stimulated). The increase in activity when co-factor is added, known as either percent stimulation or activation coefficient, is then calculated. We use the equation [(stimulated − basal)/basal] × 100 to get percent stimulation. In a deficient individual basal activity may be low and a large increase in stimulated activity should occur, resulting in a high percent stimulation.

Since these reactions are very method dependent, any laboratory which performs these analyses must define precisely the upper limit of percent stimulation expected in a non-deficient population and then select an appropriate cutoff point that suggests deficiency.

Vitamin-dependent erythrocyte enzyme assays were measured for assessment of thiamin, riboflavin, and pyridoxine status at 12 weeks gestation and at term, and compared to values found in our non-pregnant controls (Table I.I.1. and Table I.I.2.).

No effect of oral contraceptive use was seen on basal or stimulated activity or on stimulation index for any of the three enzymes tested. The values are shown for the whole control group in Table I.I.1.

ERYTHROCYTE TRANSKETOLASE AND GLUTATHIONE REDUCTASE ACTIVITY

No significant difference was found between the basal activity or stimulation index measured at 12 weeks and at term in our sample. The mean and range for our pregnant subjects at either time did not differ from non-pregnant control results. The comparisons of the 12-week values to control data are shown in Table I.I.2.

No outliers were found among our pregnant subjects for either basal transketolase activity or the stimulation index, suggesting that all had adequate thiamin status. However, one pregnant woman at term had a basal erythrocyte glutathione reductase activity of 6.7 U/G hemoglobin and a stimulation index of 98%, suggesting riboflavin deficiency. This result was excluded in determining the pregnancy reference data interval.

ERYTHROCYTE ASPARTATE AMINOTRANSFERASE ACTIVITY

Basal activity was slightly higher and the stimulation index lower in the pregnancy group compared to controls but the differences were not statistically significant. One pregnant subject had a low basal activity at 12 weeks with a stimulation index of 148%, suggesting pyridoxine deficiency, but at the term visit both basal activity and index were within the control reference range.

Clinical signs of thiamin, riboflavin, or pyridoxine deficiency were not seen in our study population. However, others have reported *biochemical* evidence of thiamin, riboflavin, or pyridoxine deficiency in 25 to 40% of middle-class pregnant women.[12,13,18,21,22] It has been suggested that moderate vitamin deficiencies can adversely affect pregnancy. Apgar scores were significantly lower in babies of low-income mothers with poor pyridoxine nutritional status shown by high erythrocyte alanine aminotransferase stimulation indices.[21] Although we found one subject with biochemical evidence of riboflavin deficiency and one with similar evidence of pyridoxine deficiency, neither had any symptoms of deficiency and both delivered healthy infants after an uncomplicated pregnancy and labor. For the other subjects, nutritional adequacy for the three vitamins was suggested by the fact that the biochemical indices, basal erythrocyte enzyme activities, and the stimulation indices for each enzyme were comparable to the control reference range for our laboratory.

OTHER VITAMINS

Vitamin B_{12}, folate, niacin, and ascorbate were not measured by our laboratory when the pregnancy study was undertaken and we do not have data for these vitamins. However, a brief commentary is provided, in view of the importance of folate deficiency in pregnancy[29-31] and the ongoing controversy surrounding the importance of nutritional factors in prevention of congenital anomalies such as neural tube defects.[32-34]

Folate or folacin are generic descriptors for the many related forms of folic acid compounds.[29-31] Adequate folate intake is essential in pregnancy to supply the requirements of increased maternal cell growth and erythropoiesis and of synthesis of fetal tissues and erythrocytes.[30,31] Although folic acid requirements like those of other nutrients impose a greater stress for women in poor socioeconomic situations, folic acid deficiency is also recognized among generally well-nourished women.[29-31]

Fetal plasma, fetal red cells, and cord blood have much higher concentrations of folic acid than maternal plasma or red cells. Folate-deficient mothers give birth to infants with normal folate levels, suggesting active transplacental transport of folate.[29,31]

There has been a substantial decrease in the Recommended Dietary Allowances for folate during pregnancy. The current recommendation of 400 μg/day is half of the 1980 recommendations.[16] The Canadian guidelines suggest 300 μg/day above non-pregnant requirements.[15]

Plasma folate concentration decreases with increasing pregnancy, mainly as a result of hemodilution so that total plasma folate is probably not decreased.[29] Hemodilution obviously does not account for the fall in erythrocyte folate which may occur. This could be interpreted as a possible indication of incipient folate deficiency.

Folate deficiency progresses through phases of low serum and then red cell folate, followed by the appearance of hypersegmented neutrophils and erythrocyte macrocytosis. Anemia indicates a late stage of deficiency.[35]

The need for vitamin B_{12}, though also essential for the rapid tissue development and growth of mother and infant, can usually be met by preexisting stores and normal dietary intake.[16] Exceptions to this include women who have been strict vegetarians for several years or those affected by chronic malabsorption involving the terminal ileum.[29] Most megaloblastic anemia in pregnancy results from deficiency of folate rather than B_{12}.

Although there is less clear identification of specific enzymes for which ascorbic acid is a co-factor, than for many of the other vitamins, ascorbate is required for hydroxylase reactions in collagen biosynthesis, catecholamine production, and amidation of peptide hormones.[36,37] Ascorbate also promotes intestinal iron absorption,[38] and probably functions as a water-soluble antioxidant similarly to the lipid-soluble antioxidant role of Vitamin E.[15]

Assessment of acute changes in ascorbic acid status is commonly done by measuring plasma or serum ascorbate, while leukocyte levels reflect longer-term nutritional status.[7] In situations of chronically low vitamin C intake, serum levels are probably quite accurate indicators.[7,39] However, it has been shown that, when ascorbic acid intake is increased, serum levels increase to a plateau of around 79.5 μmol/L or 1.4 mg/dL and do not increase further despite additional intake. Further excess ingestion is best reflected by high urinary ascorbate concentrations.

Plasma vitamin C levels decline with pregnancy due to hemodilution as well as active transport across the placenta.[15] Fetal demands for ascorbate are high. Both the Canadian and American recommendations suggest a 10-mg/day increase in ascorbate intake during pregnancy.[15,16]

Niacin is biologically important as a co-factor in its roles as nicotinamide adenine dinucleotide (NAD) and NAD phosphate (NADP). The syndrome of niacin deficiency, pellagra, characterized by gastrointestinal and neuropsychiatric dysfunction with dermatitis, is rarely seen in developing countries.[15,16] Niacin needs are increased in pregnancy because of increased energy intake.[15,16]

REFERENCES

1. **Gibson, R. S.,** Assessment of the status of thiamine, riboflavin, and niacin, in *Principles of Nutritional Assessment,* Gibson, R. S., Ed., Oxford University Press, New York, 1990, 425.
2. **Gibson, R. S.,** Assessment of vitamin B-6 status, in *Principles of Nutritional Assessment,* Gibson, R. S., Ed., Oxford University Press, New York, 1990, 445.
3. **Anon.,** Deaths associated with thiamine-deficient total parenteral nutrition, *M.M.W.R.,* 38, 43, 1989.
4. **McCormick, D. B.,** Riboflavin, in *Modern Nutrition in Health and Disease,* 7th ed., Shils, M. E. and Young, V. R., Eds., Lea & Febiger, Philadelphia, 1988, 362.

5. **McCormick, D. B.,** Two interconnected B vitamins: riboflavin and pyridoxine, *Physiol. Rev.,* 69, 1170, 1989.
6. **Sauberlich, H. E.,** Newer laboratory methods for assessing nutriture of selected B-complex vitamins, *Annu. Rev. Nutr.,* 4, 377, 1984.
7. **Gibson, R. S.,** Assessment of vitamin C status, in *Principles of Nutritional Assessment,* Gibson, R. S., Ed., Oxford University Press, New York, 1990, 413.
8. **Gibson, R. S.,** Assessment of the status of folate and vitamin B-12, in *Principles of Nutritional Assessment,* Gibson, R. S., Ed., Oxford University Press, New York, 1990, 461.
9. **Haas, R. H.,** Thiamin and the brain, *Annu. Rev. Nutr.,* 8, 483, 1988.
10. **Davis, R. E. and Icke, G.,** Clinical chemistry of thiamin, *Adv. Clin. Chem.,* 23, 93, 1983.
11. **Flannelly, G., Turner, M. J., Connolly, R., and Stronge, J. M.,** Persistent hyperemesis gravidarum complicated by Wernicke's encephalopathy, *Ir. J. Med. Sci.,* 159, 82, 1990.
12. **Heller, S., Salkeld, R. M., and Korner, W. F.,** Vitamin B1 status in pregnancy, *Am. J. Clin. Nutr.,* 27, 1221, 1974.
13. **Vir, S. C., Love, A. H. G., and Thompson, W.,** Thiamin status during pregnancy, *Int. J. Vit. Nutr. Res.,* 50, 131, 1980.
14. **Wyatt, D. T., Noetzel, M. J., and Hillman, R. E.,** Infantile beriberi presenting as subacute necrotizing encephalomyelopathy, *J. Pediatr.,* 110, 888, 1987.
15. **Scientific Review Committee, Health and Welfare Canada,** *Nutrition Recommendations,* Canadian Government Publishing Centre, Supply and Services, Ottawa, 1990.
16. **Food and Nutrition Board, National Research Council,** *Recommended Dietary Allowances,* 10th ed., National Academy Press, Washington, DC, 1989.
17. **Bates, C. J.,** Human riboflavin requirements, and metabolic consequences of deficiency in man and animals, *World Rev. Nutr. Diet.,* 50, 215, 1987.
18. **Heller, S., Salkeld, R. M., and Korner, W. F.,** Riboflavin status in pregnancy, *Am. J. Clin. Nutr.,* 27, 1225, 1974.
19. **Merrill, A. H. and Henderson, J. M.,** Diseases associated with defects in vitamin B6 metabolism or utilization, *Annu. Rev. Nutr.,* 7, 137, 1987.
20. **Rosenberg, L. E.,** Vitamin responsive inherited metabolic disorders, *Adv. Hum. Genet.,* 6, 1, 1976.
21. **Schuster, K., Bailey, L. B., and Mahan, C. S.,** Vitamin B6 status of low-income adolescent and adult pregnant women and the condition of their infants at birth, *Am. J. Clin. Nutr.,* 34, 1731, 1981.
22. **Martener-Hewes, P. M., Hunt, I. F., Murphy, N. J., Swendseid, M. E., and Settlage, R. H.,** Vitamin B-6 nutriture and plasma diamine oxidase activity in pregnant Hispanic teenagers, *Am. J. Clin. Nutr.,* 44, 907, 1986.
23. **Schuster, K., Bailey, L. B., and Mahan, C. S.,** Effect of maternal pyridoxine HCl supplementation on the Vitamin B6 status of the mother and infant and on pregnancy outcome, *J. Nutr.,* 114, 1731, 1984.
24. **Barnard, H. C., deKock, J. J., Vermaak, W. J. H., and Potgieter, G. M.,** A new perspective in the assessment of vitamin B6 nutritional status during pregnancy in humans, *J. Nutr.,* 117, 1303, 1987.
25. **Reynolds, R. D. and Leklem, J. E.,** Implications of the role of vitamin B-6 in health and disease, in *Vitamin B-6: Its Role in Health and Disease,* Reynolds, R. D. and Leklem, J. E., Eds., Alan R. Liss, New York, 1985, 481.
26. **Sahakian, V., Rouse, D., Sipes, S., Rose, N., and Niebyl, J.,** Vitamin B6 is effective therapy for nausea and vomiting of pregnancy: a randomized double-blind placebo-controlled study, *Obstet. Gynecol.,* 78, 33, 1991.
27. **Bendich, A. and Cohen, M.,** Vitamin B6 safety issues, *Ann. N.Y. Acad. Sci.,* 585, 189, 1990.

27. **Bendich, A. and Cohen, M.,** Vitamin B6 safety issues, *Ann. N.Y. Acad. Sci.,* 585, 189, 1990.
28. **Dalton, K. and Dalton, M. J. T.,** Characteristics of pyridoxine overdose neuropathy syndrome, *Acta Neurol. Scand.,* 76, 8, 1987.
29. **Metz, J.,** Folate deficiency during pregnancy and lactation, in *Folates and Cobalamins,* Zittoun, J. A. and Cooper, B. A., Eds., Springer-Verlag, Berlin, 1989, 162.
30. **Bailey, L. B. and Cerda, J. J.,** Iron and folate nutriture through the life cycle, *World Rev. Nutr. Diet.,* 56, 56, 1988.
31. **King, J., Bronstein, M. N., Fitch, W. L., and Weininger, J.,** Nutrient utilisation during pregnancy, *World Rev. Nutr. Diet.,* 52, 71, 1987.
32. **Anon.,** Evidence for an association between periconceptual use of multivitamins and neural tube defects, *Nutr. Rev.,* 48, 15, 1990.
33. **Anon.,** Evidence against an association between periconceptual use of multivitamins and neural tube defects, *Nutr. Rev.,* 48, 17, 1990.
34. **Simpson, J. L., Mills, J. L., Rhoads, G. G., Cunningham, G. C., Hoffman, H. J., and Conley, M. R.,** Vitamins, folic acid and neural tube defects: comments on investigations in the United States, *Prenat. Diag.,* 11, 641, 1991.
35. **Herbert, V.,** Making sense of laboratory tests of folate deficiency: folate requirements to sustain normality, *Am. J. Hematol.,* 26, 199, 1987.
36. **Padh, H.,** Cellular functions of ascorbic acid, *Biochem. Cell. Biol.,* 68, 1166, 1990.
37. **Moser, U. and Bendich, A.,** Vitamin C, in *Handbook of Vitamins,* 2nd ed., Machlin, L. J., Ed., Marcel Dekker, New York, 1991, 195.
38. **Hallberg, L., Brune, M., and Rossander-Hulthen, L.,** Is there a physiological role of vitamin C in iron absorption?, *Ann. N.Y. Acad. Sci.,* 498, 324, 1987.
39. **Jacob, R. A., Skalla, J. H., and Omaye, S. T.,** Biochemical indices of human vitamin C status, *Am. J. Clin. Nutr.,* 46, 818, 1987.

I.I.1. ERYTHROCYTE ENZYME ASSAYS FOR VITAMIN STATUS

Table I.I. I. Erythrocyte enzyme activity measured in non-pregnant controls, showing mean (sd) and reference interval derived from 2.5 and 97.5 percentiles.

Enzyme	Basal (U/G hemoglobin)		Stimulated (U/G hemoglobin)		Stimulation index (%)	
	Mean (sd)	Interval 2.5 97.5	Mean (sd)	Interval 2.5 97.5	Mean (sd)	Interval 2.5 97.5
Transketolase	0.76 (0.10)	0.56 0.93	0.84 (0.11)	0.62 1.04	10.0 (3.7)	0.8 18.0
Glutathione Reductase	8.53 (1.42)	6.00 11.50	11.76 (1.46)	9.10 15.20	39 (15)	10 73
Aspartate Aminotransferase	5.17 (1.31)	3.30 8.90	9.51 (1.37)	7.30 12.80	89 (20)	26 125

I.I.2. COMPARISON OF VITAMIN-DEPENDENT ERYTHROCYTE ENZYME ASSAYS

Table I.I.2. Comparison of vitamin-dependent erythrocyte enzyme assays (mean, standard error and high-low range for erythrocyte transketolase, glutathione reductase and aspartate aminotransferase activity in pregnancy, with values from non-pregnant controls.

Assay	Pregnancy Mean (sem)	low to high range found in study	Control Mean (sem)	low to high range found in study
Basal TK (U/G hemoglobin)	0.80 (0.02)	0.53 - 1.06	0.76 (0.01)	0.45 - 1.09
TK stimulation Index (%)	6.5 (0.7)	0 - 15	10 (0.4)	0.6 - 18
Basal GR (U/G hemoglobin)	9.1 (0.24)	6.3 - 11.9	8.5 (0.1)	5.8 - 11.8
GR stimulation Index (%)	32 (2)	8 - 56	39 (1)	7 - 87
Basal AST (U/G hemoglobin)	7.4 (0.5)	1.8 - 13.0	5.2 (0.1)	3.0 - 10.0
AST stimulation Index (%)	68 (4)	18 - 118	89 (2)	22 - 129

I.J. Trace Elements

EFFECT OF PREGNANCY ON TRACE ELEMENTS (ZINC, COPPER, AND SELENIUM)

Trace elements are by convention considered to be those elements normally present in the human body in milligram or microgram quantities.[1] Approximately 30% of all enzymes are metalloenzymes or require a metal ion for activation.[2] Essential trace elements, required for normal growth, development, and homeostasis include chromium, cobalt, copper, iodine, iron, manganese, molybdenum, selenium, and zinc.[3] Essential elements are of obvious importance in pregnancy.[4] Iron metabolism in pregnancy is discussed in Section I.K. Iodine is important for development and functioning of the thyroid gland. Endemic cretinism and goiter remain a problem in iodine-deficient areas of the world, although not in most developed countries.[5] Study of chromium, manganese, and molybdenum nutrition in pregnancy has been limited by the technical difficulties of measuring these ultra trace elements in biological specimens.[6,7] Copper, selenium, and zinc metabolism in pregnancy have been the focus of many nutritional studies,[4] and will be discussed below. Trace element toxicity in pregnancy, and the teratogenic importance of several elements,[8-11] will also be briefly described.

ZINC

Zinc is an essential trace element, required for normal cell replication and growth.[12-14] Growth failure, immunodeficiency, and impaired gastrointestinal and central nervous system functions are characteristic of zinc deficiency.[12] Prenatal zinc deficiency in animal models results in severe reproductive failure.[11] In pregnancies which continue to term, congenital abnormalities, particularly of the nervous system, are found.[11,12] With respect to human teratogenesis, severe human nutritional zinc deficiency has been implicated in the high incidence of congenital malformations, particularly neural tube defects in poor populations in Egypt, Iran, and Turkey.[15] However, this has not been conclusively proven.[16]

Zinc has been identified as a component or co-factor for more than 200 enzymes or proteins in different species.[12-14] Zinc may be an integral component, present in stoichiometric amounts, in zinc metalloenzymes such as carbonic anhydrase.[14] Alternatively, zinc may be a structural component, stabilizing the conformation of enzymes composed of multiple subunits, or the structure of proteins. Zinc may also perform a catalytic role, in which other elements can substitute if necessary. Zinc may stimulate or inhibit enzyme activity.[13,14]

Zinc enzymes are active in synthesis or degradation of carbohydrates, lipids, proteins, and nucleic acids.[13] Some examples are shown in Table I.J.1.

Zinc enzymes are also involved in transcription and translation of genetic information, stabilize the native structure of ribonucleic acid (RNA), and are essential for catalytic activity of RNA polymerases.[13] "Zinc fingers" are chelates of amino acids with a zinc atom which bind to control regions of deoxyribonucleic acid (DNA) and are involved in regulation of transcription.[13]

Despite the ubiquitous nature of zinc enzymes and proteins, and the critical role of zinc in cell growth and replication, it is interesting to note that few of the signs or symptoms of the zinc depletion state can be directly related to deficient activity of a specific enzyme.[14]

Acrodermatitis enteropathica is a rare autosomal recessive disorder of zinc deficiency, believed to result from an abnormality of zinc absorption. This condition is characterized by onset of infancy, failure to thrive, dermatitis, alopecia, chronic diarrhea and malabsorption, immunodeficiency, and behavioral problems.[17] Initially lethal at an early stage, treatment with diodoquin[18] and later with zinc supplementation[19] permitted survival to adult life. The reproductive capacity in diodoquin-treated women was poor. Of seven pregnancies in three-diodoquin treated women, one resulted in spontaneous abortion, one in an anencephalic pregnancy, and one achondroplastic dwarf.[20] Following zinc therapy, a woman with acrodermatitis enteropathica had two successful pregnancies.[21]

While the importance of zinc supplementation in acrodermatitis enteropathica is obvious, the issue of the importance of zinc deficiency in pregnancy uncomplicated by this inherited disorder is less clear.[16]

Zinc in Pregnancy

It is important to recognize that several factors other than nutritional deficiency can result in very low serum or plasma zinc concentration.[14,22] Zinc is a negative acute phase reactant, and levels decrease rapidly in response to stress. Diurnal variation and fasting affect zinc levels. Morning levels are higher than in specimens collected in the afternoon and are also higher following an overnight fast than in non-fasting subjects. Zinc is albumin bound, and hypoalbuminemia may lower apparent zinc concentration.[14,22] Despite these caveats, it is recognized that plasma zinc concentration in pregnancy is significantly lower than in the healthy non-pregnant woman.[16,23-27]

Our non-pregnant control range for serum zinc is comparable to that reported by others[23-26] and the relative gestational decrease in serum zinc concentration is similar to those described in other studies.[25,27] It has been suggested that the lower acceptable limit of normal for plasma zinc at 9 months gestation should be 6.1 μmol/L (40 μg/dL).[25]

In our subjects, zinc was positively correlated with serum albumin ($r = 0.6$ $p < 0.001$), suggesting that the dilutional effect on albumin may account for much of the lowered zinc level.

It should also be remembered that, although the concentration of plasma zinc may be significantly reduced due to hypoalbuminemia and plasma volume expansion, the net intravascular zinc content is actually increased. A concentration of 13 μmol/L in a plasma volume of 2.6 L approximates 34 μmol of zinc, compared to 37 μmol/L at 28 weeks gestation (9.6 μmol/L with a plasma volume of about 3.85 L).

COPPER AND CERULOPLASMIN

Copper is an essential trace element, contained in over 30 metalloenzymes,[28] of which 7, critical for normal human growth and development, are shown in Table I.J.2.

Copper concentration in plasma closely parallels the concentration of ceruloplasmin, the protein chiefly responsible for transporting copper from liver to peripheral tissues.[2,29] Copper deficiency is characterized by anemia, not responsive to supplemental iron, severe neutropenia, osteoporosis, vascular abnormalities with tortuosity of blood vessels, and central and peripheral nervous system dysfunction.[29]

Nutritional deficiency of copper has resulted from inadequately supplemented parenteral nutrition and severe malnutrition associated with prolonged diarrhea but has not been recognized in pregnancy.[30]

It is important to note that serum copper increases markedly in normal pregnancy and that a serum or plasma copper concentration within the normal range for a non-pregnant adult may indicate a disorder of copper metabolism in pregnancy. Other indices supporting a diagnosis of copper deficiency are hypochromic anemia, not responsive to iron, absolute neutropenia, and radiographic evidence of osteoporosis.[30]

The reference ranges for copper in our study and the relative increase with gestation are similar to levels reported by others.[26,31] It has been suggested that serum copper concentrations lower than 2 sd below the mean for normal pregnancies between 15 and 26 weeks occur in pregnancies resulting in anencephalic infants or spontaneous abortions.[32]

GENETIC DISORDERS OF COPPER IN PREGNANCY

Wilson's disease is an autosomal recessive disorder of copper metabolism characterized by excessive copper accumulation in various organs (liver, central nervous system, kidney) and usually by low levels of ceruloplasmin and copper in the blood.[33] Fertility is impaired in untreated women with Wilson's disease, in whom primary amenorrhea or oligomenorrhea is common.[34,35] Pregnancy may occur in women successfully treated with penicillamine[35-39] or other agents[40-42] but may be dangerous in women with hepatic dysfunction and portal hypertension.[38]

Although no major adverse maternal effects have been described from use of penicillamine during pregnancy for therapy of Wilson's disease or cystinuria, there is a concern about potential teratogenicity.[43] Generalized connective tissue disease or cutis laxa as well as other congenital malformations have been described. However, since most pregnancies on penicillamine therapy have had a successful outcome, the benefits of ongoing therapy probably outweigh the risks.[43]

The use of therapeutic zinc instead of penicillamine may prove a good alternative. High dietary intake of zinc interferes with copper absorption by stimulating metallothionein production in the intestinal mucosal cells. Zinc supplementation with zinc acetate or zinc sulfate has been successfully used in maintenance therapy of patients with Wilson's disease.[34] Since zinc is relatively non-toxic, it has been suggested as an alternative to penicillamine treatment for the pregnant Wilson's disease patient.[34,42]

SELENIUM AND GLUTATHIONE PEROXIDASE

Selenium is a component of glutathione peroxidase, an antioxidant enzyme which catabolizes hydrogen peroxide and other organic peroxides.[44-46] Other selenoproteins have been identified but their role in human metabolism is not yet known.[45,46]

Severe selenium deficiency has been associated with fatal cardiomyopathy in selenium-deficient areas of China.[47] Unsupplemented parenteral nutrition has produced profound selenium deficiency in adults, children, and infants[44,48-50] and death from cardiomyopathy has been attributed to this cause.[44,48,49] Symptoms and signs of selenium deficiency include muscle tenderness, pain and weakness, depigmentation of hair and nails, and macrocytosis.[44]

A marked geographical variation in plasma selenium levels among healthy ambulatory populations is believed to be due to extreme variations of selenium in the environment.[44] This has made investigation of selenium deficiency and toxicity infinitely more complex than that of elements such as iron, copper, and zinc. Individual patient results must be interpreted against reference ranges derived by the same laboratory from the appropriate reference population.

Selenium in Pregnancy

Plasma selenium concentration and glutathione peroxidase activity decrease in pregnancy from levels found in the non-pregnant state. In keeping with the geographical variation described above, the plasma selenium concentration in our non-pregnant female controls is higher than that reported from South Africa,[51] England,[52] and some areas of the U.S.[53,54] but similar to that reported for 6-month postpartum non-lactating mothers.[55] As most[53-57] but not all[58] reports have shown, serum or plasma selenium is decreased relative to non-pregnant female controls as early as 8 to 12 weeks of pregnancy and decreases further toward term. We found a lesser decrease in serum glutathione peroxidase activity in our pregnant women than did others.[57] Some of the decrease in selenium and glutathione peroxidase probably reflects hemodilution and is a physiological phenomenon, particularly since non-pregnant control levels are regained within a short time following birth.

However, serum selenium concentration and glutathione peroxidase activity are clearly modified by recent or ongoing selenium intake. A study in which pregnant women were given a yeast supplement, providing 100 or 200 μg selenium per day, showed that plasma selenium levels rose by 16% on 200 μg and by 4% on 100 μg.[54] Despite supplementation, a slight drop in plasma levels was seen at parturition. No clinical effects on mother or infant have been shown to result from the gestational decline in selenium.

RECOMMENDATIONS FOR ZINC, COPPER, AND SELENIUM INTAKE IN PREGNANCY

Expert committees have published recent Canadian[59] and American[60] guidelines on recommended nutrient intakes or dietary allowances for different population groups. Both recommended a daily intake of 15 mg zinc.

No recommendation for increased selenium requirement is made for the Canadian population on the basis that selenium intake from foods probably exceeds by several times the amount required for health. The American guidelines estimate that 65 μg/day would be adequate during pregnancy. Both guidelines state that insufficient data are available to define a recommended dietary allowance or recommended nutrient intake for copper, although for non-pregnant adults 1.5 to 3.0 mg/day is estimated to be safe and adequate.[60]

TOXIC TRACE ELEMENTS IN PREGNANCY

The danger of prenatal exposure to toxic metals, in organic form, trimethyl mercury or trimethyl tin, or inorganic form, lead or cadmium, first became obvious with the methyl mercury poisoning epidemics that occurred in Minamata, Japan and later in Iraq.[8–10] The fetal brain is particularly vulnerable. Asymptomatic mothers exposed to trimethylmercury had infants with profound cerebral atrophy who were severely mentally retarded.[8,61] Methyl mercury is readily transferred across the placenta to the developing fetus.[61] For most individuals, the main source of exposure to methylmercury is through consumption of fish and fish products and the chief risk occurs in communities subsisting mainly on fish.[62]

High-level lead exposure has long been recognized as an adverse factor in pregnancy outcome[63] but the effect of "low-level" blood lead in pregnancy on pre- and postnatal infant development has recently been the subject of several large epidemiological studies.[64-66] The 1991 revision of the statement by the Centers for Disease Control on *Preventing Lead Poisoning in Young Children* decreased the blood lead concentration deemed indicative of excessive exposure from 1.21 μmol/L to 0.48 μmol/L (from 25 μg/dl down to 10 μg/dl).[66]

Since our study population is living in an urban environment, we were concerned about environmental lead exposure. We measured blood lead concentration in our subjects during the first and third trimester and at the 6th postpartum week to determine whether there was any alteration over the months of pregnancy.

Blood lead concentration did not change between the first trimester and the time of birth. The mean (sd) blood lead level was 0.20 (0.08) μmol/L at the 12 week and term visit and 0.26 (0.11) μmol/L at the 6th week post partum. These values were not significantly different to our non-pregnant control mean of 0.19 (0.08) μmol/L. These values are approximately half those found in studies from Kentucky in 1982[67] and similar to recent maternal blood lead values reported from Roskilde in Denmark.[68] The levels in our population probably reflect decreased in environmental lead pollution since leaded gasoline was banned in Canada. This indicates that a high lead burden is not a problem in our local population. However, recent studies have shown that, even at blood lead concentrations previously considered low, i.e., 0.48 μmol/L (10 μg/dL), adverse effects on both mother and infant may be noted.[69] These include a doubling of the risk for minor congenital malformations (hemangiomas, syndactyly, polydactyly, hydrocele, undescended testicles) in women with blood lead of 0.34 to 0.48 μmol/L compared to those with a blood lead of <0.05 μmol/L. Higher systolic blood pressure and a higher incidence of pregnancy-induced hypertension were also associated with higher levels of maternal blood lead.[69]

Teratogenicity has been shown for elements such as cadmium and nickel in various animal species but no evidence for teratogenicity has been shown in humans.[10,70]

REFERENCES

1. **Frieden, E.,** A survey of the essential biochemical elements, in *Biochemistry of the Essential Ultratrace Elements,* Frieden, E., Ed., Plenum Press, New York, 1984, 1.
2. **Frieden, E.,** Perspectives on copper biochemistry, *Clin. Physiol. Biochem.,* 4, 11, 1986.
3. **Hambidge, K. M.,** Trace elements in human nutrition, in *Nutrition in Pediatrics,* Walker, W. A. and Watkins, J. B., Eds., Little, Brown, Boston, 1985, 17.
4. **Campbell, D. M.,** Trace element needs in pregnancy, *Proc. Nutr. Soc.,* 47, 45, 1988.
5. **Delange, F. M.,** Anomalies in physical and intellectual development associated with severe goitre, in *Towards the Eradication of Endemic Goitre, Cretinism and Iodine Deficiency,* Dunn, J. T., Pretell, E. A., Daza, C. H., and Viteri, F., Eds., Pan American Health Organization, Washington, DC, 1986, 49.
6. **Versieck, J.,** Trace elements in human body fluids and tissues, *Crit. Rev. Clin. Lab. Sci.,* 22, 97, 1985.

7. **Delves, H. T.,** Assessment of trace element status, *Clin. Endocrinol. Metab.*, 14, 725, 1985.
8. **Annau, Z.,** Organometals and brain development, *Prog. Brain. Res.*, 73, 295, 1988.
9. **Koos, B. J. and Longo, L. D.,** Mercury toxicity in the pregnant woman, fetus and newborn infant, *Am. J. Obstet. Gynecol.*, 126, 390, 1976.
10. **Chang, L. W. and Wade, P. R.,** Prenatal and neonatal toxicology and pathology of heavy metals, *Adv. Pharmacol. Chemother.*, 17, 195, 1980.
11. **Hurley, W. L. and Doane, R. M.,** Recent developments in the roles of vitamins and minerals in reproduction, *J. Dairy Sci.*, 72, 784, 1989.
12. **Sandstead, H.,** Zinc: essentiality for brain development and function, *Nutr. Rev.*, 43, 129, 1985.
13. **Vallee, B. L.,** Zinc, biochemistry, physiology, toxicology and clinical pathology, *Biofactors,* 1, 31, 1988.
14. **Cunnane, S. C.,** Enzymes, in *Zinc: Clinical and Biochemical Significance,* Cunnane, S. C., Ed., CRC Press, Boca Raton, FL, 1988, 79.
15. **Sever, L. E. and Emanuel, I.,** Is there a connection between maternal zinc deficiency and congenital malformations of the central nervous system in man, *Teratology,* 7, 117, 1973.
16. **Apgar, J.,** Zinc and reproduction, *Annu. Rev. Nutr.*, 5, 43, 1985.
17. **Neldner, K. H., Hambidge, K. M., and Walravens, P. A.,** Acrodermatitis enteropathica, *Int. J. Dermatol.*, 17, 330, 1978.
18. **Dillaha, C. J., Lorinez, A. L., and Aavik, O. R.,** Acrodermatitis enteropathica: a review of the literature and report of a case successfully treated with diodoquin, *JAMA,* 152, 509, 1953.
19. **Moynahan, E. J.,** Acrodermatitis enteropathica: a lethal inherited zinc deficiency disorder, *Lancet,* ii, 339, 1974.
20. **Hambidge, K. M., Neldner, K. H., and Walravens, P. A.,** Zinc, acrodermatitis enteropathica and congenital malformations, *Lancet,* I, 57, 1973.
21. **Benton, D. P., Jackson, M. J., and Young, A.,** Two pregnancies in a patient with acrodermatitis enteropathica treated with zinc sulfate, *Lancet,* ii, 500, 1981.
22. **Pilch, S. M. and Senti, F. R.,** Analysis of zinc data from the Second National Health and Nutrition Survey (NHANES II), *J. Nutr.*, 115, 1393, 1985.
23. **McMichael, A. J., Dreosti, I. E., Gibson, G. T., Hartshorne, J. M., Buckley, R. A., and Colley, D. P.,** A prospective study of serial maternal serum zinc levels and pregnancy outcome, *Early Hum. Dev.*, 7, 59, 1982.
24. **Swanson, C. A. and King, J. C.,** Reduced serum zinc concentration during pregnancy, *Obstet. Gynecol.*, 62, 313, 1983.
25. **Hambidge, K. M., Krebs, N. F., Jacobs, M. A., Favier, A., Guyette, L., and Ikle, D. N.,** Zinc nutritional status during pregnancy: a longitudinal study, *Am. J. Clin. Nutr.*, 37, 429, 1983.
26. **Tuttle, S., Aggett, P. J., Campbell, D., and MacGillivray, I.,** Zinc and copper nutrition in human pregnancy: a longitudinal study in normal primigravidae and in primigravidae at risk of delivering a growth retarded baby, *Am. J. Clin. Nutr.*, 41, 1032, 1985.
27. **Mukherjee, M. D., Sandstead, H. H., Ratnaparkhi, V., Johnson, L. K., Milne, D. B., and Stelling, H. P.,** Maternal zinc, iron, folic acid, and protein nutriture and outcome of human pregnancy, *Am. J. Clin. Nutr.*, 40, 496, 1984.
28. **Harris, E. D.,** Copper transport: an overview, *Proc. Soc. Exp. Biol.*, 196, 130, 1991.
29. **Danks, D. M.,** Copper deficiency in humans, *Annu. Rev. Nutr.*, 8, 235, 1988.
30. **Hambidge, K. M. and Mauer, A. M.,** Trace elements, in *Laboratory Indices of Nutritional Status in Pregnancy,* National Research Council Committee on Nutrition of the Mother and Preschool Child, National Academy of Sciences, Washington, DC, 1978, 157.

31. **Vir, S. C., Love, A. H. G., and Thompson, W.,** Serum and hair concentrations of copper during pregnancy, *Am. J. Clin. Nutr.*, 34, 2382, 1981.
32. **Buamah, P. K., Russell, M., Milford-Ward, A., Taylor, P., and Roberts, D. F.,** Serum copper concentration significantly less in abnormal pregnancies, *Clin. Chem.*, 30, 1676, 1984.
33. **Zbella, E. A. and Gleicher, N.,** Disorders of metals and metalloproteins, in *Principles of Medical Therapy in Pregnancy,* Gleicher, N., Ed., Plenum Press, New York, 1985.
34. **Brewer, G. J. and Yuzbasiyan-Gurkan, V.,** Wilson's disease: an update with emphasis on new approaches to treatment, *Dig. Dis.*, 7, 178, 1989.
35. **Walsche, J. M.,** Pregnancy in Wilson's disease, *Q. J. Med.*, 46, 73, 1977.
36. **Toaff, R., Toaff, M. E., Peyser, M. R., and Streifler, M.,** Hepatolenticular degeneration (Wilson's disease) and pregnancy, *Obstet. Gynecol. Surv.*, 32, 497, 1977.
37. **Albukerk, J. N.,** Wilson's disease and pregnancy, a case report, *Fertil. Steril.*, 24, 494, 1973.
38. **Scheinberg, I. H. and Sternlieb, I.,** Pregnancy in penicillamine treated patients with Wilson's disease, *N. Engl. J. Med.*, 293, 1300, 1975.
39. **Dupont, P., Irion, O., and Beguin, F.,** Pregnancy in a patient with treated Wilson's disease, *Am. J. Obstet. Gynecol.*, 163, 1527, 1990.
40. **Dreifuss, F. E. and McKinney, W. M.,** Wilson's disease (hepatolenticular degeneration) and pregnancy, *JAMA,* 195, 960, 1966.
41. **Walshe, J. M.,** The management of Wilson's disease treated with trientine, *Q. J. Med.*, 225, 81, 1986.
42. **Lao, T. T. H., Chin, R. K. H., Cockram, C. S., and Leung, N. W. Y.,** Pregnancy in a woman with Wilson's disease treated with zinc, *Asia Oceania J. Obstet. Gynecol.*, 14, 167, 1988.
43. **Rosa, F. W.,** Teratogen update: penicillamine, *Teratology,* 33, 127, 1986.
44. **Lockitch, G.,** Selenium: clinical significance and analytical concepts, *Crit. Rev. Clin. Lab. Sci.*, 27, 483, 1989.
45. **Burk, R. F.,** Recent developments in trace element metabolism and function: newer roles of selenium in nutrition, *J. Nutr.*, 119, 1051, 1989.
46. **Sunde, R. A.,** Molecular biology of selenoproteins, *Annu. Rev. Nutr.*, 10, 451, 1990.
47. **Yang, G., Ge, K., Chen, J., and Chen, X.,** Selenium-related endemic diseases and the daily selenium requirement of humans, *World Rev. Nutr. Diet.*, 55, 98, 1988.
48. **Johnson, R. A., Baker, S. S., Fallon, J. T., Maynard, E. P., III, Ruskin, J. N., Wen, Z., Ge, K., and Cohen, H. J.,** An occidental case of cardiomyopathy and selenium deficiency, *N. Engl. J. Med.*, 304, 1210, 1981.
49. **Lockitch, G., Taylor, G. P., Wong, L. T. K., Davidson, A. G. F., Dison, P. J., Riddell, D., and Massing, B.,** Cardiomyopathy associated with nonendemic selenium deficiency in a Caucasian adolescent, *Am. J. Clin. Nutr.*, 52, 572, 1990.
50. **Lockitch, G., Jacobson, B. E., Quigley, G., Dison, P., and Pendray, M. R.,** Selenium deficiency in low birthweight neonates: an unrecognized problem, *J. Pediatr.*, 114, 865, 1989.
51. **Heese, H. de V., Lawrence, M. A., Dempster, W. S., and Pocock, F.,** Reference concentrations of serum selenium and manganese in healthy nulliparas, *S.A.M.J.*, 73, 163, 1988.
52. **Hinks, L. J., Ogilvy-Stuart, A., Hambidge, K. M., and Walker, V.,** Maternal zinc and selenium status in pregnancies with a neural tube defect or elevated plasma alpha-fetoprotein, *Br. J. Obstet. Gynaecol.*, 96, 61, 1989.
53. **Swanson, C. A., Reamer, D. C., Veillon, C., King, J. C., and Levander, O. A.,** Quantitative and qualitative aspects of selenium utilization in pregnant and nonpregnant women: an application of stable isotope methodology, *Am. J. Clin. Nutr.*, 38, 169, 1983.

54. **Butler, J. A. and Whanger, P. D.,** Dietary selenium requirements of pregnant women and their infants, in *Selenium in Biology and Medicine,* Combs, G. F., Spallholz, J. E., Levander, O. A., and Oldfield, J. E., Eds., Van Nostrand Reinhold, New York, 1987, 688.
55. **Levander, O. A., Moser, P. B., and Morris, V. C.,** Dietary selenium intake and selenium concentrations of plasma, erythrocytes, and breast milk in pregnant and post-partum lactating and nonlactating women, *Am. J. Clin. Nutr.,* 46, 694, 1987.
56. **Behne, D. and Wolters, W.,** Selenium content and glutathione peroxidase activity in the plasma and erythrocytes of non-pregnant and pregnant women, *J. Clin. Chem. Clin. Biochem.,* 17, 133, 1979.
57. **Butler, J. A., Whanger, P. D., and Tripp, M. J.,** Blood selenium and glutathione peroxidase activity in pregnant women: comparative assays in primates and other animals, *Am. J. Clin. Nutr.,* 36, 15, 1982.
58. **Kundu, N., Parke, P., Petersen, L. P., Palmer, I. S., and Olson, O.,** Distribution of serum selenium, copper and zinc in normal human pregnancy, *Arch. Env. Health,* 40, 268, 1985.
59. **Scientific Review Committee, Health and Welfare Canada,** *Nutrition Recommendations,* Canadian Government Publishing Centre, Supply and Services, Ottawa, 1990.
60. **Food and Nutrition Board, National Research Council,** *Recommended Dietary Allowances,* 10th ed., National Academy Press, Washington, DC, 1989.
61. **Takeuchi, T.,** Pathology of Minamata disease, *Acta Pathol. Jpn.,* 32, 73, 1982.
62. **Inskip, M. J. and Piotrowski, J. K.,** Review of the health effects of methylmercury, *J. Appl. Toxicol.,* 5, 113, 1985.
63. **McMichael, A. J., Baghurst, P. A., Wigg, N. R., Vimpani, G. V., Robertson, E. F., and Roberts, R. J.,** Port Pirie Cohort Study: environmental exposure to lead and children's abilities at the age of four years, *N. Engl. J. Med.,* 319, 468, 1988.
64. **Bellinger, D., Leviton, A., Waternaux, C., Needleman, H., and Rabinowitz, M.,** Longitudinal analyses of prenatal and postnatal lead exposure and early cognitive development, *N. Engl. J. Med.,* 316, 1037, 1987.
65. **Dietrich, K. N., Krafft, K. M., Bornschein, R. L., Hammond, P. B., Berger, O., Succop, P. A., and Bier, M.,** Low-level fetal lead exposure effect on neurobehavioral development in early infancy, *Pediatrics,* 80, 721, 1987.
66. **Centers for Disease Control,** *Preventing Lead Poisoning in Young Children,* U.S. Department of Health and Human Services, Atlanta, 1991, 1.
67. **Angell, N. F. and Lavery, J. P.,** The relationship of blood lead to obstetric outcome, *Am. J. Obstet. Gynecol.,* 142, 40, 1982.
68. **Milman, N., Christensen, J. M., and Ibsen, K. K.,** Blood lead and erythrocyte zinc protoporphyrin in mothers and newborn infants, *Eur. J. Pediatr.,* 147, 71, 1988.
69. **Rabinowitz, M., Lead and pregnancy,** *Birth,* 15, 236, 1988.
70. **Sager, P. R., Clarkson, T. E., and Nordberg, G. F.,** Reproductive and developmental toxicity, in *Handbook on the Toxicology of Metals,* Friberg, G., Nordberg, G. F., and Vouk, V. B., Eds., Elsevier, Amsterdam, 1986, 391.

I.J.1. SOME ZINC METALLOENZYMES OR METALLOPROTEINS IMPORTANT IN HUMAN PHYSIOLOGY

Table I.J.1. Zinc metalloenzymes or metalloproteins important in human physiology

Enzyme or protein	Function
Alkaline phosphatase	heterogeneous enzymes catalysing non-specific hydrolysis of phosphate esters. Important in bone calcification, liver and intestinal transport.
Carbonic anhydrase	catalyses reversible hydration of CO_2 to H_2CO_3
Alcohol dehydrogenase	catalyses interconversion of ethanol and other primary alcohols with corresponding aldehydes.
Delta-aminolevulinic acid dehydratase	catalyses porphobilinogen synthesis from two molecules of delta-aminolevulinic acid
α-D-Mannosidase	deficient in autosomal recessive disorder of mannosidosis
Angiotensin-converting enzyme	activates angiotensin I to angiotensin II, potent vasopressor
Fructose-1,6-diphosphatase	important in carbohydrate metabolism
Superoxide dismutase	dismutation of superoxide radicals
Metallothionein	cysteine-rich proteins with high content of metals including zinc, cadmium, copper, iron - abundant in kidney and liver, function not clearly defined

I.J.2. SOME (COPPER) METALLOENZYMES IMPORTANT IN HUMAN PHYSIOLOGY

Table I.J.2. Copper metalloenzymes important in human physiology

Enzyme	Function	Effect of Deficiency
Ceruloplasmin	copper transport from liver, iron transfer to transferrin, synthesis of biogenic amines, cell growth, antioxidation, angiogenesis	copper deficiency syndromes (anemia, bone cartilage, vascular and central nervous system abnormalities)
Cytochrome C-oxidase	electron transport	deficient energy production, temperature instability
Superoxide dismutase	dismutation of superoxide radicals	oxidative tissue injury
Lysyl oxidase	cross-linking of collagen and elastin	cartilage, vessel, bone and joint abnormalities
Tyrosinase	melanin production	lack of pigment in hair and skin
Amine oxidases	oxidation of biogenic amines	increased histamine, urticaria
Dopamine-ß-hydroxylase	catecholamine metabolism	central and peripheral nervous system malfunction

I.J.3. ZINC

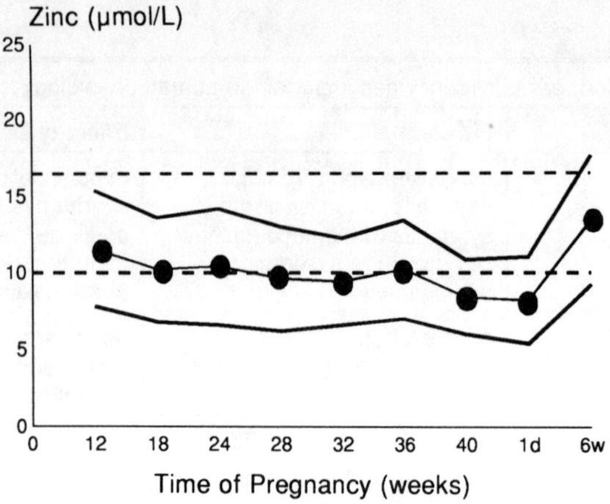

Figure I.J.3. Gestational changes in zinc concentration

Table I.J.3. Zinc (μmol/L)			
Time of Gestation	Reference Range	Mean (sd)	Percentage of Control mean
Non Pregnant Control	10.0 - 16.5	13.0 (1.7)	
12 weeks	7.8 - 15.4	11.6 (1.9)	89
18 weeks	6.8 - 13.6	10.2 (1.7)	78
24 weeks	6.6 - 14.2	10.4 (1.9)	80
28 weeks	6.2 - 13.0	9.6 (1.7)	74
32 weeks	6.6 - 12.2	9.4 (1.4)	72
36 weeks	7.0 - 13.4	10.2 (1.6)	78
Term	6.0 - 10.8	8.4 (1.2)	65
1 day pp	5.4 - 11.0	8.2 (1.4)	63
6 week pp	9.2 - 17.6	13.4 (2.1)	103

I.J.4. COPPER

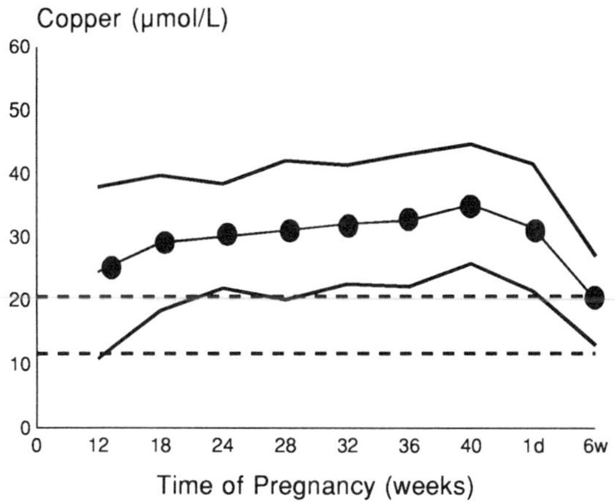

Figure I.J.4. Gestational changes in copper concentration.

Table I.J.4. Copper (μmol/L)

Time of Gestation	Reference Range	Mean (sd)	Percentage of Control mean
Non Pregnant Control	11.7 - 20.5 #	15.4 (2.3)	
12 weeks	10.8 - 37.9	24.4 (6.8)	158
18 weeks	18.3 - 39.7	29.0 (5.3)	188
24 weeks	21.8 - 38.4	30.0 (4.1)	195
28 weeks	20.0 - 42.0	31.0 (5.5)	201
32 weeks	22.4 - 41.3	32.0 (4.7)	208
36 weeks	22.0 - 43.0	32.5 (5.2)	211
Term	25.6 - 44.5	35.0 (4.7)	227
1 day pp	21.3 - 41.4	31.4 (5.0)	204
6 week pp	12.9 - 26.8	19.9 (3.5)	129

Non-pregnant control reference range - data from OC users excluded

I.J.5. SELENIUM

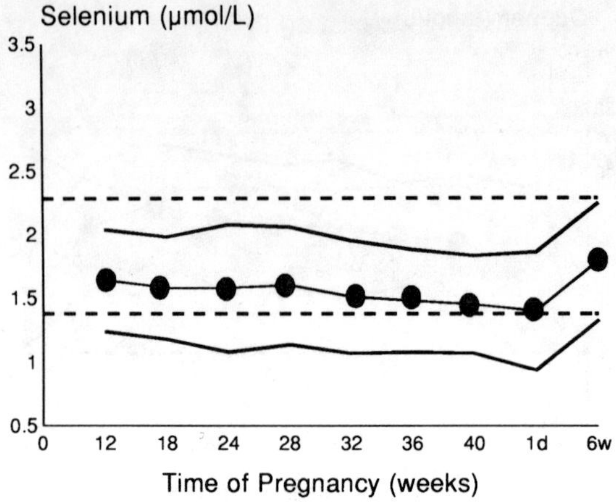

Figure I.J.5. Gestational changes in selenium concentration.

Table I.J.5. Selenium (µmol/L)			
Time of Gestation	Reference Range	Mean (sd)	Percentage of Control mean
Non Pregnant Control	1.38 - 2.29 #	1.71 (0.22)	
12 weeks	1.24 - 2.04	1.66 (0.19)	97
18 weeks	1.18 - 1.98	1.58 (0.20)	92
24 weeks	1.08 - 2.08	1.58 (0.25)	92
28 weeks	1.14 - 2.06	1.60 (0.23)	94
32 weeks	1.07 - 1.95	1.51 (0.22)	88
36 weeks	1.08 - 1.88	1.48 (0.20)	87
Term	1.07 - 1.83	1.45 (0.19)	85
1 day pp	0.94 - 1.86	1.40 (0.23)	82
6 week pp	1.33 - 2.25	1.80 (0.23)	105

Non-pregnant control reference range - data from OC users excluded

I.J.6. GLUTATHIONE PEROXIDASE

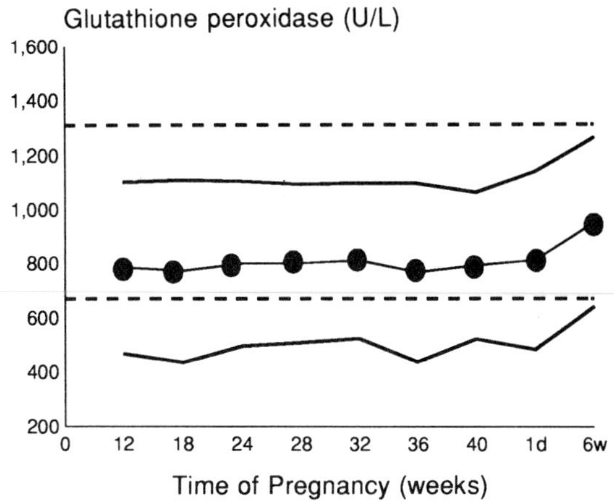

Figure I.J.6. Gestational changes in glutathione peroxidase concentration.

Table I.J.6. Glutathione Peroxidase (U/L) (serum)

Time of Gestation	Reference Range	Mean (sd)	Percentage of Control mean
Non Pregnant Control	670 - 1312	992 (172)	
12 weeks	467 - 1103	785 (159)	79
18 weeks	434 - 1111	772 (169)	78
24 weeks	494 - 1106	800 (153)	81
28 weeks	507 - 1095	801 (147)	81
32 weeks	522 - 1100	811 (168)	82
36 weeks	435 - 1098	767 (201)	77
Term	521 - 1064	792 (135)	80
1 day pp	482 - 1141	812 (165)	82
6 week pp	639 - 1265	952 (156)	96

I.K. Indices of Iron Metabolism

EFFECT OF PREGNANCY ON IRON METABOLISM

Approximately 1000 mg of additional iron are required in pregnancy for expansion of the erythrocyte mass, synthesis of fetal and placental tissue, and replacement of basal iron loss from normal exfoliation of cells.[1] The estimated iron requirement for expansion of erythrocyte mass is about 450 mg.[1] Third trimester fetal accretion of iron occurs at about 1.6 to 2.0 mg/kg/day.[2] A newborn term infant contains about 280 to 300 mg iron[1,3] or about 75 mg/kg.[2] The placenta and umbilical cord contain between 50 and 75 mg iron.[1,3]

Available sources of iron are preexisting maternal iron stores and dietary iron. Cessation of iron loss due to menstruation compensates for the increased needs in the first trimester.[4] Iron absorption increases with increasing gestation from about 12.5% of dietary intake to 25% in the second and third trimesters of pregnancy.[5] However, the diet alone may not supply adequate amounts of iron, particularly if stores are low.[4,5]

It has been estimated that about 8 mg of iron are contained in iron stores for each 1 μg/L of serum ferritin.[6] The mean (range) of serum ferritin in non-pregnant women has been variously reported by different authors to be 23 (4 to 145),[7] 25 (4 to 144),[8] and 34 (9 to 125) μg/L.[9] In our non-pregnant controls the mean and range were 27 (4 to 105) μg/L.

On the basis of these various studies, iron stores in non-pregnant women could range from about 184 to 464 mg.[3] This implies that an additional 530 to 800 mg must be obtained from the diet over the course of pregnancy, i.e., about 2 to 3 mg/day in excess of normal daily requirements.[4] Since about 25% of dietary iron is absorbed during pregnancy, an increment in daily intake of about 15 mg should be adequate for most women.[4] In those women with minimal preexisting stores, almost all the iron must be obtained from oral intake.[4,5] Since the iron content of habitual North American diets is unlikely to provide this,[4] a daily iron supplement is recommended for the second and third trimester,[4,5] or earlier if evidence of red cell microcytosis develops.

It is important to realize that interactions between elements may occur and interfere with intestinal absorption.[10] Women receiving an iron supplement had significantly lower serum zinc concentration compared to a control group not receiving iron.[10] Both zinc and manganese have been shown to interfere with iron absorption, either by direct competition for intestinal absorptive pathways or by intraluminal interaction with dietary ligands.[11]

IRON SUPPLEMENTATION IN PREGNANCY

The need for iron supplementation in pregnancy has been controversial.[12-14] It is recognized that in a healthy pregnancy there is a physiological decrease in hemoglobin concentration, hematocrit, and erythrocyte count resulting from expansion of plasma volume.[15-17] However, although hemoglobin concentration and erythrocyte counts are lower, the increase in volume of distribution actually results in an overall increase in erythrocyte mass.[15]

Adverse pregnancy outcome has been associated with both decreased and increased hemoglobin concentration and hematocrit.[16-18] Anemia in pregnancy has been associated with reduced infant birth weight and increased placental size.[3,19,20] The resultant increase in placental:fetal weight ratio is like that seen in smokers and in pregnancies at high altitudes,[21] suggesting that the cause is lowered oxygen tension.[3]

Body iron content is reduced in fetuses of anemic mothers. A study of aborted fetuses showed that the iron content in fetuses of mothers with hemoglobins of >120 g/L was 75 mg/kg body weight compared with 63 mg/kg if maternal hemoglobin was <90 g/L.[22]

Conversely, it has also been shown that relatively high hemoglobin concentration and hematocrit, associated with failure of adequate expansion of plasma volume, may result in intrauterine growth retardation and fetal death.[16-18] It was postulated that increased viscosity predisposes to intervillous thrombosis and infarction, thus impairing uteroplacental circulation.[14,17,23] This could account for the increase in low birth weight infants born to supplemented mothers.[14,23]

The National Collaborative Perinatal Project in which more than 50,000 pregnancies were studied revealed that the optimal maternal hemoglobin and hematocrit were, respectively, between 110 to 120 g/L and 0.33 to 0.35.[18] Patients with a hemoglobin <90 or >130 g/L or a hematocrit <0.29 or >0.39 had increased risk of fetal death, premature birth, or low birth weight infants.[18]

The controversy over iron requirements in pregnancy and the need for supplementation has generally focused on the issues of anemia and the optimal hemoglobin or hematocrit for pregnancy outcome.[15-20] However, iron, as an essential trace element, is a component of other heme and non-heme proteins and metalloenzyme systems, as well as a co-factor for other enzymes.[24,25]

Iron deficiency is associated with a decrease in myoglobin, important for oxygen transport and storage in muscle.[24,25] Also decreased are iron-containing mitochondrial enzymes of the electron transport chain.[25] Iron deficiency results in alterations in dopaminergic and serotonergic neurotransmission.[25] There is thus increasing interest in non-hematological aspects of iron deficiency, such as impairment in work performance[26] and impaired behavioral and cognitive functions.[27]

The issue of iron supplementation in pregnancy remains of interest.[13,14] Recent conclusions have varied from a recommendation that all pregnant women receive iron supplementation from 16 weeks[12] to the recommendation that only those in whom "true iron deficiency anemia" is diagnosed should be treated.[14]

Iron deficiency is extremely common in women of child-bearing age, even in developed countries.[1,27-30] Using bone marrow examination or serum ferritin to indicate low or absent iron stores, minimal iron storage was estimated to be a problem in 30% of Swedish and Finnish women,[27,28] 16% of continental French women but 40% of migrant workers,[29] and 10% of women aged 20 to 40 in the U.S.[30]

Among our healthy non-pregnant controls, 20/121 (17%) had ferritin concentrations $\leq$12 μg/L and 12/21 (10%) were $\leq$6 μg/L. Thus, a significant number of apparently healthy women who become pregnant may lack sufficient iron stores to support the requirements of a full-term pregnancy.

It may be most appropriate to predicate iron supplementation in pregnancy on the basis of a hemoglobin, mean cell volume, and serum ferritin measured early in pregnancy. If these are all normal, iron supplementation is not immediately required. If there is evidence of iron depletion early in pregnancy, iron supplementation should probably be started at once. Patients should be retested for iron depletion in the late second trimester on this basis.

EFFECT OF PREGNANCY ON PARAMETERS OF IRON METABOLISM

The differentiation between dilutional anemia and the anemia of iron deficiency is complex since both the indices of iron metabolism and erythrocyte indices (Section II.) are altered by the physiological changes of normal pregnancy.

Serum iron concentration decreases with increasing duration of pregnancy but the total iron content of the blood is minimally changed because of expanded plasma volume.[15] The concentration of serum iron shows both diurnal fluctuation[31] and a negative response in the acute phase reaction.[32] A fasting early morning level of $\leq$10.7 μmol/L has been suggested to indicate significant iron depletion in the non-pregnant state.[33] Total iron concentration is a relatively insensitive indicator of body iron depletion and is probably less useful in assessing iron status in pregnancy than ferritin.[23,33,34]

Transferrin increases by about 2.4 times due to estrogen-stimulated hepatic protein synthesis[3] (Section I.F.). Transferrin saturation accordingly decreases even during healthy pregnancy from the mean of about 0.33 in the non-pregnant state to around 0.25.

Ferritin is an acute phase reactant and increases in infection, malignancy, rheumatoid arthritis, liver disease, and hyperthyroidism.[6] Ferritin levels in hemochromatosis are several thousand-fold higher than in the normal population.[6] A ferritin of $\leq$12 μg/L has been considered to indicate significant iron depletion in non-pregnant subjects.[33]

In pregnancy, studies have consistently demonstrated that ferritin decreases from about the 12th to the 15th week until about 28 to 32 weeks as iron stores are used for expansion of erythrocyte mass.[1,3,23,34] This fall occurs whether or not iron supplements are given, although the levels are lower in unsupplemented women.[23,34] Few iron-supplemented women will have ferritin levels <10 μg/L, although this is commonly seen in pregnant women who do not receive supplemental iron.[23] In the last trimester of pregnancy, despite a major increase in transplacental iron transport to the fetus,[1,3] ferritin levels are relatively stable, suggesting that increased iron absorption can supply increased fetal needs.

Erythrocyte protoporphyrin increases when heme synthesis is disrupted by iron deficiency.[33,34] Experimental iron depletion in non-pregnant women over a 2- to 3-month period indicated that neither erythrocyte protoporphyrin by total extraction nor zinc protoporphyrin by hematofluorometry was as sensitive as ferritin estimations in identifying iron depletion.[33] However, in a group of pregnant women, zinc protoporphyrin was more useful than iron concentration, transferrin saturation, or ferritin as a means to predict development of third trimester anemia (defined as a hemoglobin <115 g/L) in pregnant women, given oral iron supplementation.[34]

USE OF THESE TESTS IN PREGNANCY

Hemoglobin and mean corpuscular volume should be routinely measured as part of prenatal bloodwork. Should either be low, the serum ferritin should be measured, and supplementation started. In iron-deficient patients, hemoglobin measurements should be repeated at 28 and 36 weeks and the iron supplements adjusted depending on the results. The ranges in I.K.1 to I.K.2 include data from the whole pregnancy group whether on iron supplements or not.

REFERENCES

1. **Hallberg, L.,** Iron balance in pregnancy, in *Vitamins and Minerals in Pregnancy and Lactation,* Berger, H., Ed., Raven Press, New York, 1988, 115.
2. **Ehrenkranz, R. E.,** Mineral needs of the very low birth-weight infant, *Sem. Perinatol.,* 13, 142, 1989.
3. **Rosso, P.,** Maternal metabolism of vitamins and minerals, in *Nutrition and Metabolism in Pregnancy: Mother and Fetus,* Rosso, P., Ed., Oxford University Press, New York, 1990, 66.
4. **National Research Council, Food and Nutrition Board,** *Recommended Dietary Allowances,* 10th ed., National Academy Press, Washington, DC, 1989, 195.
5. **Health and Welfare Canada, Scientific Review Committee,** Iron, in *Nutrition Recommendations,* Canadian Government Publishing Centre, Ottawa, 1990, 141.
6. **Worwood, M.,** Serum ferritin, *Crit. Rev. Clin. Lab. Sci.,* 10, 171, 1979.

7. **Valberg, L. S., Sorbie, J., Ludwig, T., and Pelletier, O.,** Serum ferritin and the iron status of Canadians, *C.M.A.J.*, 114, 417, 1976.
8. **Luxton, A. W. and Walker, W. H. C.,** A radioimmunoassay for serum ferritin, *Clin. Chem.*, 23, 683, 1977.
9. **Cook, J. D. and Lipschitz, D. A.,** Serum ferritin as a measure of iron stores in normal subjects, *Am. J. Clin. Nutr.*, 27, 681, 1974.
10. **Bloxam, D. L., Williams, N. R., Waskett. R. J., Pattinso-Green, P. M., Morarji, Y., and Stewart, S. G.,** Maternal zinc during oral iron supplementation in pregnancy: a preliminary study, *Clin. Sci.*, 76, 59, 1989.
11. **Rossander-Hulten, L., Brune, M., Sandstrom, B., Lonnerdal, B., and Hallberg, L.,** Competitive inhibition of iron absorption by manganese and zinc in humans, *Am. J. Clin. Nutr.*, 54, 152, 1991.
12. **Letsky, E. A.,** Blood volume, hematinics, anemia, in *Medical Disorders in Obstetric Practice,* 2nd ed., de Swiet, M., Ed., Blackwell Scientific, Oxford, 1989, 48.
13. **Reece, J., Donovan, P. R., and Pellett, A. Y.,** Iron supplementation in pregnancy; testing a new clinical protocol, *J. Am. Diet. Assoc.*, 87, 1682, 1987.
14. **Hytten, F. E.,** Nutritional requirements in pregnancy: what should the pregnant women be eating?, *Midwifery,* 6, 93, 1990.
15. **Hytten, F. E.,** Blood volume changes in normal pregnancy, *Clin. Haematol.*, 14, 601, 1985.
16. **Goodlin, R. C., Dobry, C. A., Anderson, J. C., Woods, R. E., and Quaife, M.,** Clinical signs of normal plasma volume expansion during pregnancy, *Am. J. Obstet. Gynecol.*, 124, 1001, 1983.
17. **Koller, O.,** The clinical significance of hemodilution during pregnancy, *Obstet. Gynecol. Surv.*, 37, 649, 1982.
18. **Garn, S. M., Ridella, S. A., Petzold, A. S., and Falkner, F.,** Maternal hematological levels and pregnancy outcomes, *Sem. Perinatol.*, 5, 155, 1981.
19. **Beischer, N. A., Sivasamboo, R., Vohra, S. S., and Reid, S.,** Placental hypertrophy in severe pregnancy anemia, *J. Obstet. Gynaecol. Br. Commonw.*, 77, 398, 1970.
20. **Harrison, K. A. and Ibeziako, P. A.,** Maternal anemia and fetal birthweight, *J. Obstet. Gynaecol. Br. Commonw.*, 80, 798, 1973.
21. **Wingerd, J., Christianson, R., Lovitt, W. V., and Schoen, E. J.,** Placental ratio in black and white women: relation to smoking and anemia, *Am. J. Obstet. Gynecol.*, 124, 671, 1976.
22. **Ahmad, S. H., Amir, M., Ansari, Z., and Ahmed, K. N.,** Influence of maternal iron deficiency on the fetal total body iron, *Ind. Pediatr.*, 20, 643, 1983.
23. **Mahomed, K. and Hytten, F. E.,** Iron and folate supplementation in pregnancy, in *Effective Care in Pregnancy and Childbirth,* Chalmers, I., Enkin, M., and Kierse, M., Eds., Oxford University Press, Oxford, 1989.
24. **Dallman, P. R.,** Biochemical basis for the manifestations of iron deficiency, *Annu. Rev. Nutr.*, 6, 13, 1986.
25. **Lozoff, B.,** Behavioural alterations in iron deficiency, *Adv. Pediatr.*, 35, 331, 1988.
26. **Groner, J. A., Holtzman, N. A., Charnery, E., and Mellits, E. D.,** A randomized trial of oral iron on tests of short term memory and attention span in young pregnant women, *J. Adol. Health Care,* 7, 44, 1986.
27. **Rybo, E.,** Diagnosis of iron deficiency, *Scand. J. Hematol.*, Suppl. 43, 34, 1985.
28. **Takkunen, H.,** Iron deficiency in the Finnish adult population, *Scand. J. Hematol. Suppl.*, 25, 1, 1976.
29. **Hercberg, S., Galan, P., Soutre, Y., Devanlay, M., and Dupin, H.,** Prevalence of iron deficiency during pregnancy in a French area, *Nutr. Rep. Int.*, 32, 719, 1985.
30. **Expert Working Group,** Summary of a report on the assessment of iron nutritional status of the United States population, *Am. J. Clin. Nutr.*, 42, 1318, 1985.

31. **Long, R., Delaney, K. K., and Siegel, L.,** Diurnal variation of serum iron in normal individuals, *Clin. Chem.,* 24, 842, 1978.
32. **Fleck, A.,** Clinical and nutritional aspects of changes in acute phase proteins during inflammation, *Proc. Nutr. Soc.,* 48, 347, 1989.
33. **Milne, D. B., Gallagher, S. K., and Nielsen, F. H.,** Response of various indices of iron status to acute iron depletion produced in menstruating women by low iron intake and phlebotomy, *Clin. Chem.,* 36, 487, 1990.
34. **Schifman, R., Thomasson, J. E., and Evers, J. M.,** Red blood cell zinc protoporphyrin testing for iron-deficiency anemia in pregnancy, *Am. J. Obstet. Gynecol.,* 157, 304, 1987.

I.K.1. IRON

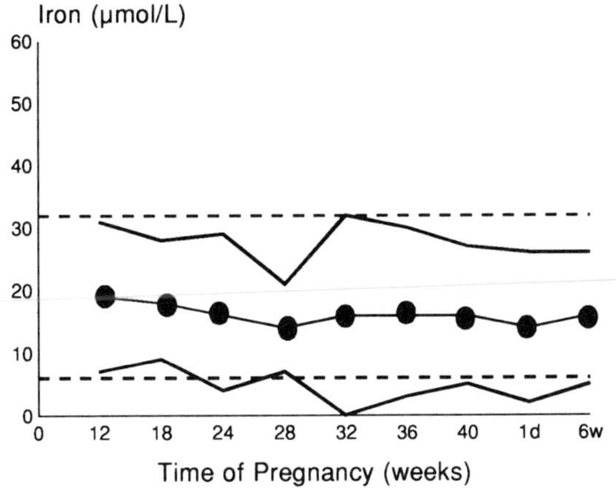

Figure I.K.1. Gestational changes in iron concentration.

Table I.K.1. Iron (µmol/L)			
Time of Gestation	Reference Range	Mean (sd)	Percentage of Control mean
Non Pregnant Control	6 - 32 #	17 (7)	
12 weeks	7 - 31	19 (6)	112
18 weeks	9 - 28	18 (5)	106
24 weeks	4 - 29	16 (6)	94
28 weeks	7 - 21	14 (4)	82
32 weeks	0 - 32	16 (8)	94
36 weeks	3 - 30	16 (7)	94
Term	5 - 27	16 (5)	94
1 day pp	2 - 26	14 (6)	82
6 week pp	5 - 26	16 (5)	94

Non-pregnant control reference range - data from OC users excluded

I.K.2. IRON-BINDING CAPACITY

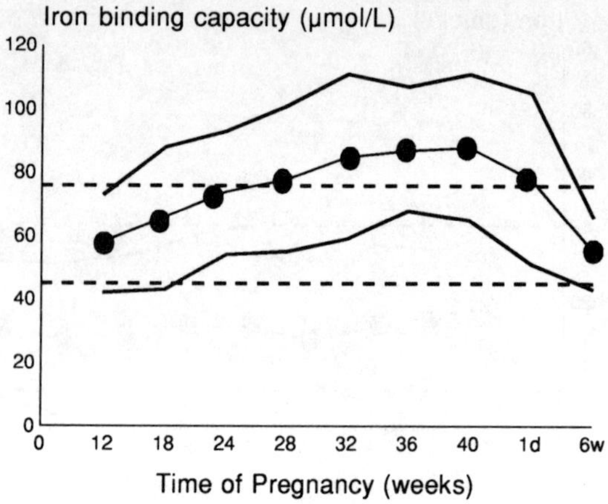

Figure I.K.2. Gestational changes in iron binding capacity

Table I.K.2. Iron binding capacity (µmol/L)			
Time of Gestation	Reference Range	Mean (sd)	Percentage of Control mean
Non Pregnant Control	45 - 76 #	61 (8)	
12 weeks	42 - 73	58 (8)	95
18 weeks	43 - 88	66 (12)	108
24 weeks	54 - 93	74 (10)	121
28 weeks	55 - 101	79 (11)	129
32 weeks	59 - 111	85 (13)	139
36 weeks	68 - 107	87 (10)	142
Term	65 - 111	88 (11)	144
1 day pp	51 - 105	78 (13)	128
6 week pp	43 - 66	55 (6)	90

Non-pregnant control reference range - data from OC users excluded

I.K.3. TRANSFERRIN SATURATION

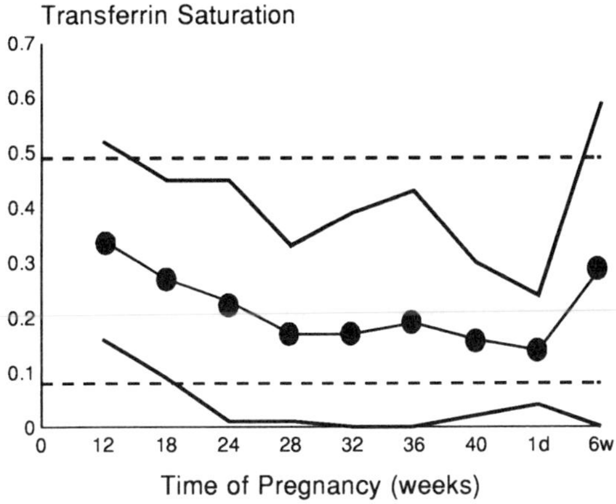

Figure I.K.3. Gestational changes in transferrin saturation concentration.

Table I.K.3. Transferrin saturation. This is a derived value.
Transferrin saturation = [iron (μmol/L) / {transferrin (g/L) x 23.1}]

Time of Gestation	Reference Range	Mean (sd)	Percentage of Control mean
Non Pregnant Control	0.08 - 0.49	0.25 (0.11)	
12 weeks	0.16 - 0.52	0.34 (0.09)	136
18 weeks	0.09 - 0.45	0.27 (0.09)	108
24 weeks	0.01 - 0.45	0.23 (0.11)	92
28 weeks	0.01 - 0.33	0.17 (0.08)	68
32 weeks	0 - 0.39	0.17 (0.11)	68
36 weeks	0 - 0.43	0.19 (0.12)	76
Term	0.02 - 0.30	0.16 (0.07)	64
1 day pp	0.04 - 0.24	0.14 (0.05)	56
6 week pp	0 - 0.59	0.29 (0.15)	116

I.K.4. ZINC PROTOPORPHYRIN

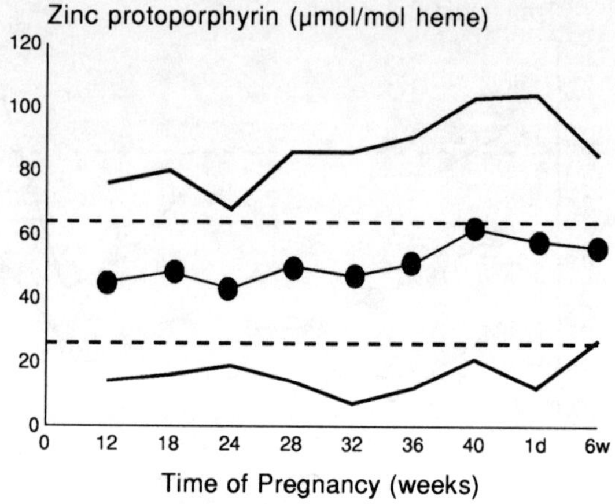

Figure I.K.4. Gestational changes in zinc protoporphyrin concentration.

Table I.K.4. Zinc protoporphyrin (μmol/mol heme)

Time of Gestation	Reference Range	Mean (sd)	Percentage of Control mean
Non Pregnant Control	26 - 64	43 (12)	
12 weeks	14 - 76	46 (15)	107
18 weeks	16 - 80	48 (16)	112
24 weeks	19 - 68	43 (12)	100
28 weeks	14 - 86	50 (18)	116
32 weeks	7 - 86	47 (20)	109
36 weeks	12 - 91	51 (20)	119
Term	21 - 103	62 (21)	144
1 day pp	12 - 104	58 (23)	135
6 week pp	27 - 85	56 (14)	130

I.K.5 FERRITIN

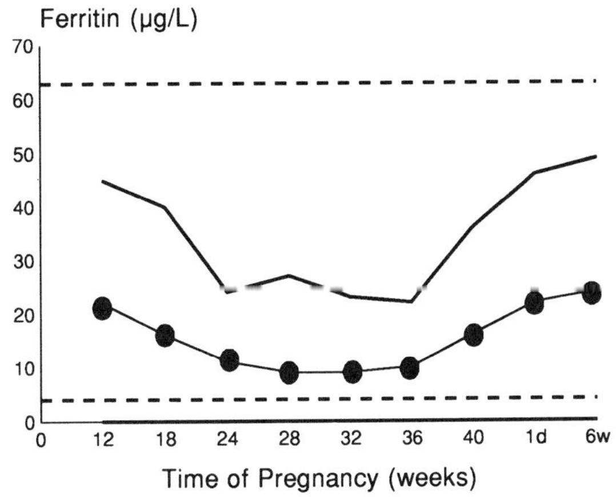

Figure I.K.5. Gestational changes in ferritin concentration.

Table I.K.5. Ferritin (μg/L)

Time of Gestation	Reference Range	Mean (sd)	Percentage of Control mean
Non Pregnant Control	4 - 63	27 (7)	
12 weeks	0 - 45	22 (12)	81
18 weeks	0 - 40	16 (12)	59
24 weeks	0 - 24	10 (7)	37
28 weeks	0 - 27	9 (9)	33
32 weeks	0 - 23	9 (7)	33
36 weeks	0 - 22	10 (6)	37
Term	0 - 36	16 (10)	59
1 day pp	0 - 46	22 (12)	81
6 week pp	0 - 49	24 (13)	95

I.L. Thyroid Hormones, Thyroxine-Binding Globulin, and Cortisol

EFFECT OF PREGNANCY ON THYROID HORMONES

Although normal pregnancy is considered to be a euthyroid state, the hormonal changes and metabolic demands of pregnancy result in significant changes in thyroid function.

Serum thyroxine (T4) and triiodothyronine (T3) concentrations increase gradually throughout the first half of pregnancy and remain constant during the latter half of pregnancy. This increase is due to increased thyroxine-binding globulin (TBG) production.

During the first trimester, free T4 does not significantly change. During the second and third trimester, the mean free T4 falls by about 25%. This has been shown to be independent of albumin, TBG, and prealbumin concentration.[1-3]

Thyrotropin (thyroid-stimulating hormone, TSH) changes reciprocally with free T4. TSH is usually low in the first trimester and returns to non-pregnant levels by the second trimester. High human chorionic gonadotropin (HCG) concentration in early pregnancy is thought to have TSH-like activity *in vivo*. Hence, to maintain normal free T4, hypothalamus-pituitary-thyroid feedback control may require less TSH. In later trimesters, HCG concentrations decline, allowing TSH to resume its usual regulatory potency.[4]

Maternal TSH, T4, and T3 do not cross the placenta and fetal thyroid function is autonomous. However, both thyroid stimulatory and inhibitory immunoglobulins may cross the placenta, resulting in different manifestations of thyroid disease in the offspring of sequential pregnancies.[5]

HYPERTHYROIDISM

The introduction of sensitive radioimmunoassays for the measurement of TSH allows discrimination between TSH levels found in euthyroid and hyperthyroid patients. In hyperthyroidism, pituitary TSH secretion is suppressed and falls below the detection limit of the assay. Free T4, total T4, and T3 are significantly increased. The Free Thyroxine Index (FTI) and the resin T3 uptake may be inappropriately normal during pregnancy complicated by hyperthyroidism and should not be relied upon.[6]

HYPOTHYROIDISM

The most sensitive indicator of primary hypothyroidism in pregnancy is an elevated TSH concentration. The physiologic changes seen during pregnancy make the other thyroid function tests less reliable, although they still provide supportive evidence.[6,7]

OTHER CAUSES OF ABNORMAL THYROID FUNCTION TESTS DURING PREGNANCY

Higher frequencies of increased total and free T4 have been noted with severe hyperemesis gravidarum or morning sickness of early pregnancy. Most of the cases are transient and levels return to normal by the second trimester. Thyroid microsomal antibodies and TSH are usually normal. Serum HCG has been proposed as the cause of the thyrotropic activity. Thyrotoxicosis is also associated with hydatidiform mole and choriocarcinoma. In such cases, where thyrotoxicosis is present, the TSH levels are undetectable and the serum HCG levels are markedly elevated.[6]

THYROID DISEASE IN PREGNANCY

HYPERTHYROIDISM

Although hyperthyroidism affects less than 0.2% of pregnancies, diagnosis of this condition in pregnancy is important.[8,9] Untreated hyperthyroidism is associated with premature labor and fetal mortality.[9] Thyrotoxic crisis can be lethal to mother and infant. Fewer maternal and fetal complications occur when hyperthyroidism is recognized and treated prior to conception.[9] Maternal hyperthyroidism during the period of organogenesis has been associated with an increased incidence of major nonthyroidal congenital malformations.[10]

The symptoms of thyrotoxicosis in pregnancy are similar to the thyrotoxic syndrome in the non-pregnant adult. Symptoms include tremor, heat intolerance, irritability, insomnia, palpitations, and tachycardia. Thyroid crisis with cerebral dysfunction, hyperthermia, gastrointestinal hypermotility, and heart failure may complicate 2% of treated thyrotoxicosis cases but may occur in as many as 20% of untreated patients.[9]

Women who have had thyroid ablation for hyperthyroidism and are euthyroid may still have infants affected by intrauterine or neonatal thyrotoxicosis,[5,11] since thyroid-stimulating antibodies may persist for years after thyroidectomy or radioactive iodine treatment.

HYPOTHYROIDISM

Most severely hypothyroid women are anovulatory[7] and untreated hypothyroidism is rarely diagnosed in pregnancy. In overt hypothyroidism, both maternal complications and perinatal morbidity and mortality are common.

Women who were hypothyroid in pregnancy had a higher incidence of abortions, stillbirths, prematurity, and congenital abnormalities than euthyroid women.[7,12] Fetal death occurred in 12% of pregnancies.[7] The incidence of congenital anomaly in untreated hypothyroidism was four times higher than the expected rate and was also higher than expected in treated hypothyroid patients.[12]

Normal pregnancy can occur in maternal hypothyroidism. An uncomplicated full-term pregnancy has been described in a woman in whom hypothyroidism was not diagnosed until late in pregnancy.[13,14] If hypothyroidism is diagnosed, replacement therapy with thyroid hormone is required.

THYROID ANTIBODIES IN PREGNANCY

It is important to remember, as mentioned above, that maternal thyroid antibodies can cause thyroid dysfunction in the fetus and newborn infant.[5] Maternal thyroid-stimulating immunoglobulins can cross the placenta and cause neonatal or intrauterine thyrotoxicosis in both treated or untreated maternal Graves disease.[5]

Transient neonatal hypothyroidism has been caused by transplacental passage of maternal anti-TSH antibodies.[15-17] These growth-blocking antibodies against the TSH growth receptor were found in the mothers of almost half the infants with sporadic congenital hypothyroidism detected in a screening program.[18]

EFFECT OF PREGNANCY ON ADRENAL HORMONES

During the normal course of pregnancy, a state of physiological hypercortisolism occurs. Total and free cortisol in plasma begin to rise early in the second trimester and by term are about twice that of non-pregnant women. At all times, the diurnal rhythm is maintained as reflected in the significantly higher values for cortisol in the morning than in the afternoon.[19] Cortisol-binding globulin (CBG) concentration also begins to rise during the second trimester and remains approximately twice non-pregnant values until term.

It has been suggested that the increase in total cortisol concentration is due to the increase in CBG concentrations. This would not, however, explain the rise in *free* cortisol levels. The adrenal gland appears to be more responsive to adrenocorticotropin (ACTH) during pregnancy with a greater rise in plasma cortisol concentration for a given dose of ACTH.[20-22]

Corticotropin-releasing hormone (CRH) and ACTH also increase during pregnancy due to placental production but do not begin to rise until the end of the second trimester, *after* the rise in free cortisol. In addition, CRH and ACTH secretion do not have a diurnal variation and fall rapidly postpartum, unlike cortisol which takes up to a week to return to normal.[21] For these reasons, the increase in free cortisol during pregnancy cannot be entirely explained by CRH and ACTH stimulation.

ADRENAL DISEASE DURING PREGNANCY

CUSHING'S SYNDROME

Cushing's syndrome rarely presents during pregnancy because amenorrhea and infertility are characteristic of the disease.[23] Despite this, over 60 cases have been reported in the literature.[23-25] The diagnosis is often difficult because weight gain, pigmented striae, edema, hypertension, and diabetes frequently occur during pregnancy. The most common cause of Cushing's syndrome, over all, is pituitary-dependent adrenal hyperplasia, with the remaining cases split among benign adrenal adenoma, adrenal carcinoma, and ectopic ACTH syndrome. In contrast, the majority of cases of Cushing's syndrome during pregnancy are benign adrenal adenomas, with pituitary-dependent adrenal hyperplasia being the second leading cause.[25] Regardless of the cause, complications are frequent and include gestational diabetes, severe hypertension, premature labor, and stillbirths.[26]

The laboratory diagnosis of Cushing's syndrome in pregnancy is complicated by the changes in adrenal function during normal pregnancy. Patients with Cushing's syndrome typically have elevated plasma cortisol concentrations *without* diurnal variation. This is in contrast to normal pregnancy where cortisol is elevated but diurnal variation is maintained. Both the low dose (2 mg) and high dose (8 mg) dexamethasone suppression tests, with plasma and urinary measurements of cortisol, are reliable during pregnancy.[26] Pregnant women with normal adrenal function will suppress with both these tests. In response to the high-dose test, urinary cortisol should fall to less than 50% of basal values in the majority of patients with pituitary-dependent disease. Most patients with ectopic tumors, adrenal adenomas, and carcinomas show no suppression. The overnight 1-mg dexamethasone suppression test (followed next morning by a single measurement of plasma cortisol) is often abnormal in normal pregnancy and therefore should not be used as a screening test for adrenal disease during pregnancy.[27]

ADRENOCORTICAL INSUFFICIENCY

Adrenocortical insufficiency, acute or chronic, is also rare in pregnancy. Most patients present with weakness, fatigue, increased pigmentation, weight loss, and persistent nausea and vomiting. In many cases, the symptoms were attributed to the pregnancy itself and the diagnosis was made for the first time after adrenal crisis developed during the puerperium. More than half the reported cases, who carried their pregnancies to term, developed acute adrenal failure postpartum. This tendency to tolerate adrenal insufficiency during pregnancy and then develop adrenal crisis postpartum may be related to the diuresis that follows delivery.[26]

Plasma cortisol concentrations may be within the normal non-pregnancy range in patients with Addison's disease. The diagnosis rests on the lack of rise in the plasma cortisol concentration after Cortrosyn (ACTH) infusion.

HYPERALDOSTERONISM

Only a few cases of hyperaldosteronism during pregnancy have been reported in the literature.[28,29] As in the non-pregnant state, hypertension, hypokalemia, and urinary potassium wasting are usually present. Aldosterone progressively rises throughout normal pregnancy and is approximately tenfold higher than non-pregnant values by term.[29] For this reason, urinary aldosterone measurements are difficult to interpret during pregnancy.

PHEOCHROMOCYTOMA

Pheochromocytoma is a very rare cause of hypertension during pregnancy. Symptomatology usually consists of paroxysmal hypertension, sweating, headache, and tachycardia but may closely mimic severe preeclampsia.[30] Its significance is that, left undiagnosed, pheochromocytoma is associated with maternal and fetal mortality rates greater than 50%. In normal pregnancy, urinary excretion of catecholamines, vanillylmandelic acid (VMA), and metanephrine do not change. The reliability of urinary catecholamine measurement to diagnose pheochromocytoma during pregnancy is similar to that of the non-pregnant state.[31]

REFERENCES

1. **Ballabio, M., Poshyachinda, M., and Ekins, R. P.,** Pregnancy induced changes in thyroid function: role of HCG as putative regulator of maternal thyroid, *J. Clin. Endocrinol. Metab.*, 73, 824, 1991.
2. **Ball, R., Freedman, D. B., Holmes, J. C., Midgley, J. E., M., and Sheehan, C. P.,** Low normal concentration of free thyroxin in serum in late pregnancy: physiological fact, not technical artefact, *Clin. Chem.*, 35, 1891, 1989.
3. **Guillaume, J., Schussler, G. C., and Goldman, J.,** Components of the total serum thyroid hormone concentration during pregnancy: high free thyroxine and blunted thyrotropin (TSH) response to TSH-releasing hormone in the first trimester, *J. Clin. Endocrinol. Metab.*, 60, 678, 1985.
4. **Chan, B. Y.,** Serum thyrotrophin concentration measured by sensitive assays in normal pregnancy, *Br. J. Obstet. Gynaecol.*, 95, 1332, 1988.
5. **Fort, P., Lifshitz, F., Pugliese, M., and Klein, I.,** Neonatal thyroid disease: differential expression in three successive offspring, *J. Clin. Endocrinol. Metab.*, 66, 645, 1988.
6. **Becks, G. P. and Burrow, G. N.,** Thyroid disease and pregnancy, *Med. Clin. North Am.*, 75, 121, 1991.
7. **Davis, L. E., Leveno, K. J., and Cunningham, F. G.,** Hypothyroidism complicating pregnancy, *Obstet. Gynecol.*, 72, 108, 1988.
8. **Burrow, G. N.,** The management of thyrotoxicosis in pregnancy, *N. Engl. J. Med.*, 313, 562, 1985.
9. **Davis, L. E., Lucas, M. J., Hankins, G. D. V., Roark, M. L., and Cunningham, F. G.,** Thyrotoxicosis complicating pregnancy, *Am. J. Obstet. Gynecol.*, 160, 63, 1989.
10. **Momotani, N., Ito, K., Hamada, N., Ban, Y., Nishikawa, Y., and Mimura, T.,** Maternal hyperthyroidism and congenital malformation in the offpsring, *Clin. Endocrinol.*, 20, 695, 1984.
11. **Cove, D. H. and Johnston, P.,** Fetal hyperthyroidism: experience of treatment in four siblings, *Lancet*, 1, 430, 1985.
12. **Potter, J. D.,** Hypothyroidism and reproductive failure, *Surg. Gynecol. Obstet.*, 150, 251, 1980.
13. **Montoro, M., Collea, J. V., Frasier, S. D., and Mestman, J. H.,** Successful outcome of pregnancy in women with hypothyroidism, *Ann. Intern. Med.*, 94, 31, 1981.

14. **Kennedy, A. L. and Montgomery, D. A. D.,** Hypothyroidism in pregnancy, *Br. J. Obstet. Gynaecol.,* 85, 225, 1978.
15. **Matsurra, N., Yamada, Y., Nohara, Y., Konishi, J., Kasagi, K., Endo, K., Kohima, H., and Wataya, K.,** Familial neonatal transient hypothyroidism due to maternal TSH-binding inhibitor immunoglobulins, *N. Engl. J. Med.,* 303, 738, 1980.
16. **Iseki, M., Shimizu, M., Oikawa, T., Hojo, H., Arikawa, K., Ichikawa, Y., Momotani, N., and Ito, K.,** Sequential serum measurements of thyrotropin binding inhibitor immunoglobulin G in transient familial neonatal hypothyroidism, *J. Clin. Endocrinol. Metab.,* 57, 384, 1983.
17. **Konishi, J., Iida, Y., Kasagi, K., Misaki, T., Nakashima, T., Endo, K., Mori, T., Shinpo, S., Nohara, Y., and Matsuura, N.,** Primary myxedema with thyrotrophin-binding inhibitor immunoglobulins, *Ann. Intern. Med.,* 103, 26, 1985.
18. **Van Der Gaag, R. D., Drexhage, H. A., and Dussault, J. H.,** Role of maternal immunoglobulins blocking TSH-induced thyroid growth in sporadic forms of congenital hypothyroidism, *Lancet,* i, 246, 1985.
19. **Meulenberg, P. M. and Hofman, J. A.,** Differences between concentrations of salivary cortisol and cortisone and of free cortisol and cortisone in plasma during pregnancy and postpartum, *Clin. Chem.,* 36, 70, 1990.
20. **Scott, E. M., McGarrigle, H. H. G., and Lachelin, G. C. L.,** The increase in plasma and saliva cortisol levels in pregnancy is not due to the increase in corticosteroid binding globulin levels, *J. Clin. Endocrinol. Metab.,* 71, 639, 1990.
21. **Okamoto, E., Takagi, T., Makino, T., Sata, H., Iwaka, I., Nishino, E., Mitsudo, N., Sugita, N., Otsuki, Y., and Tanizawa, O.,** Immunoreactive corticotropin-releasing hormone, adrenocorticotropin and cortisol in human plasma during pregnancy and delivery and postpartum, *Horm. Metab. Res.,* 21, 566, 1989.
22. **Baxter, J. D. and Tyrrell, J. B.,** The adrenal cortex, in *Endocrinology and Metabolism,* 2nd ed., Felig, P., Baxter, J. D., Broadus, A. E., and Frohman, L. A., Eds., McGraw-Hill, New York, 1987, 511.
23. **Aron, D. C., Schnall, A. M., and Sheeler, L. R.,** Cushing's syndrome and pregnancy, *Am. J. Obstet. Gynecol.,* 162, 244, 1990.
24. **Mazor, M., Leiberman, J. R., Korenblum, R., and Katz, M.,** Pregnancy complicated by Cushing's syndrome, *Arch. Gynecol. Obstet.,* 241, 191, 1987.
25. **Buesch, M. A., McClamrock, H. D., and Adashi, E. Y.,** Cushing's syndrome in pregnancy, *Obstet. Gynecol.,* 79, 130, 1992.
26. **Burrow, G. N.,** Pituitary and adrenal disorders, *Medical Complications during Pregnancy,* Burrow, G. N. and Ferris, T. F., Eds., W. B. Saunders, Philadelphia, 1988, 254.
27. **Smith, R., Owens, P. C., Brinsmead, M. W., Singh, B., and Hall, C.,** The nonsuppressibility of plasma cortisol persists after pregnancy, *Horm. Metab. Res.,* 19, 41, 1987.
28. **Hammond, T. G., Buchanan, J. D., Scoggins, B. A., Thatcher, R., and Whitworth, J. A.,** Primary hyperaldosteronism in pregnancy, *Aust. N.Z. J. Med.,* 12, 537, 1982.
29. **Lotgering, F. K., Derkx, F. M. H., and Wallenburg, H. C. S.,** Primary hyperaldosteronism in pregnancy, *Am. J. Obstet. Gynecol.,* 155, 986, 1986.
30. **Ellison, G. T., Mansberger, J. A., and Mansberger, A. R.,** Malignant recurrent pheochromocytoma during pregnancy: case report and review of the literature, *Surgery,* 103, 484, 1988.
31. **Freier, D. T., Eckhauser, F. E., and Harrison, T. S.,** Pheochromocytoma. A persistently problematic and still potentially lethal disease, *Arch. Surg.,* 115, 388, 1980.

I.L.1. THYROXINE

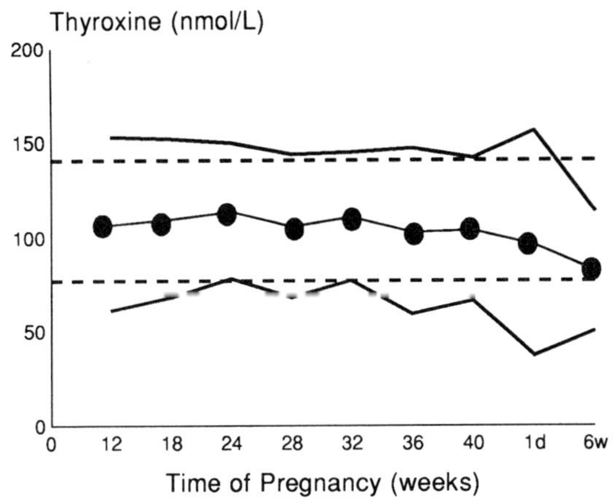

Figure I.L.1. Gestational changes in thyroxine concentration

Table I.L.1. Thyroxine (nmol/L)			
Time of Gestation	Reference Range	Mean (sd)	Percentage of Control mean
Non Pregnant Control	77 - 141 #	104 (17)	
12 weeks	61 - 153	107 (23)	103
18 weeks	68 - 152	110 (21)	106
24 weeks	78 - 150	114 (18)	110
28 weeks	68 - 144	106 (19)	102
32 weeks	77 - 145	111 (17)	107
36 weeks	59 - 147	103 (22)	99
Term	66 - 142	104 (19)	100
1 day pp	37 - 156	96 (30)	92
6 week pp	50 - 114	82 (16)	79

Non-pregnant control reference range - data from OC users excluded

I.L.2. TRIIODOTHYRONINE

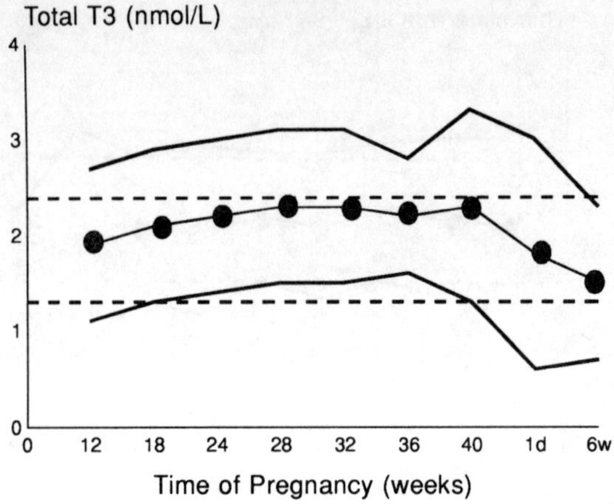

Figure I.L.2. Gestational changes in triiodothyronine concentration.

Table I.L.2. Triiodothyronine (nmol/L)

Time of Gestation	Reference Range	Mean (sd)	Percentage of Control mean
Non Pregnant Control	1.3 - 2.4 #	1.9 (0.3)	
12 weeks	1.1 - 2.7	1.9 (0.4)	100
18 weeks	1.3 - 2.9	2.1 (0.4)	111
24 weeks	1.4 - 3.0	2.2 (0.4)	116
28 weeks	1.5 - 3.1	2.3 (0.4)	121
32 weeks	1.5 - 3.1	2.3 (0.4)	121
36 weeks	1.6 - 2.8	2.2 (0.3)	116
Term	1.3 - 3.3	2.3 (0.5)	121
1 day pp	0.6 - 3.0	1.8 (0.6)	95
6 week pp	0.7 - 2.3	1.5 (0.4)	79

Non-pregnant control reference range - data from OC users excluded

I.L.3. FREE THYROXINE

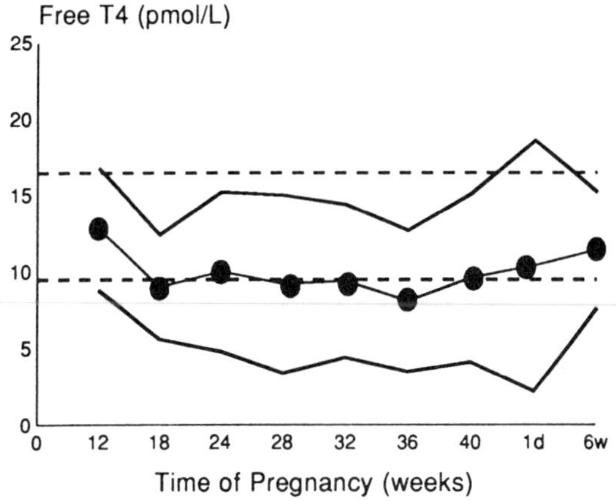

Figure I.L.3. Gestational changes in free thyroxine concentration.

Table I.L.3. Free Thyroxine (pmol/L)			
Time of Gestation	Reference Range	Mean (sd)	Percentage of Control mean
Non Pregnant Control	9.5 - 16.5	13.0 (1.9)	
12 weeks	8.8 - 16.8	12.8 (2.0)	98
18 weeks	5.6 - 12.4	9.0 (1.7)	69
24 weeks	4.8 - 15.2	10.0 (2.6)	77
28 weeks	3.4 - 15.0	9.2 (2.9)	71
32 weeks	4.4 - 14.4	9.4 (2.5)	72
36 weeks	3.5 - 12.7	8.1 (2.3)	62
Term	4.1 - 15.1	9.6 (2.7)	74
1 day pp	2.2 - 18.6	10.4 (4.1)	80
6 week pp	7.6 - 15.3	11.4 (1.9)	88

I.L.4. THYROXINE-BINDING GLOBULIN

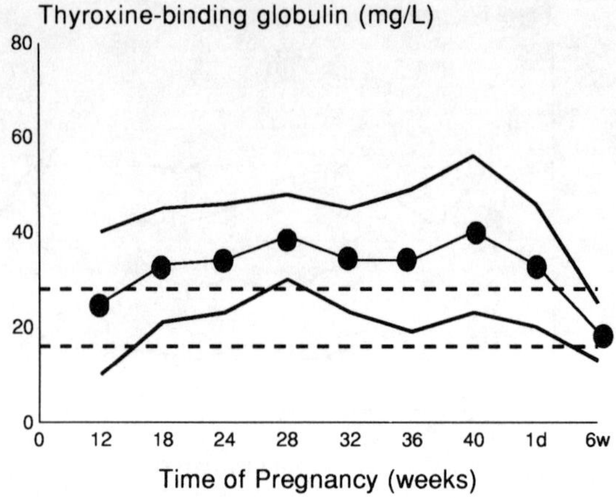

Figure I.L.4. Gestational changes in thyroxine-binding globulin concentration.

Table I.L.4. Thyroxine-binding globulin (mg/L)			
Time of Gestation	Reference Range	Mean (sd)	Percentage of Control mean
Non Pregnant Control	16 - 28 #	22 (3)	
12 weeks	10 - 40	25 (7)	114
18 weeks	21 - 45	33 (6)	150
24 weeks	23 - 46	34 (6)	155
28 weeks	30 - 48	39 (4)	177
32 weeks	23 - 45	34 (6)	155
36 weeks	19 - 49	34 (8)	155
Term	23 - 56	40 (8)	182
1 day pp	20 - 46	33 (7)	150
6 week pp	13 - 25	19 (3)	86

Non-pregnant control reference range - data from OC users excluded

I.L.5. THYROID STIMULATING HORMONE

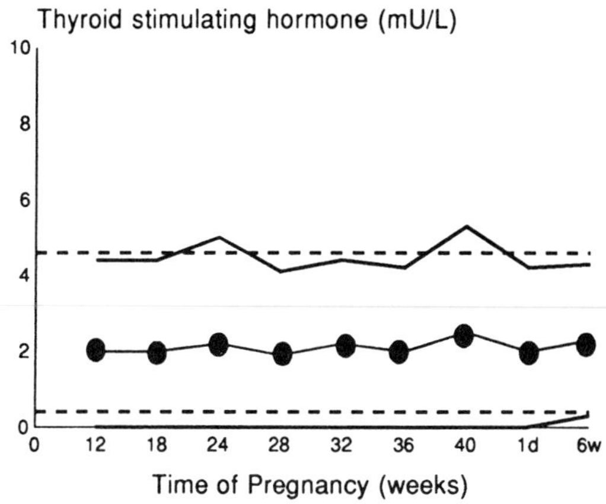

Figure I.L.5. Gestational changes in thyroid stimulating hormone concentration.

Table I.L.5. Thyroid stimulating hormone (mU/L)			
Time of Gestation	Reference Range	Mean (sd)	Percentage of Control mean
Non Pregnant Control	0.4 - 4.6	1.8 (1.0)	
12 weeks	0 - 4.4	2.0 (1.4)	111
18 weeks	0 - 4.4	2.0 (1.4)	111
24 weeks	0 - 5.0	2.2 (1.4)	122
28 weeks	0 - 4.1	1.9 (1.1)	106
32 weeks	0 - 4.4	2.2 (1.1)	122
36 weeks	0 - 4.2	2.0 (1.1)	111
Term	0 - 5.3	2.5 (1.4)	139
1 day pp	0 - 4.2	2.0 (1.2)	111
6 week pp	0.3 - 4.3	2.3 (1.0)	128

I.L.6. CORTISOL

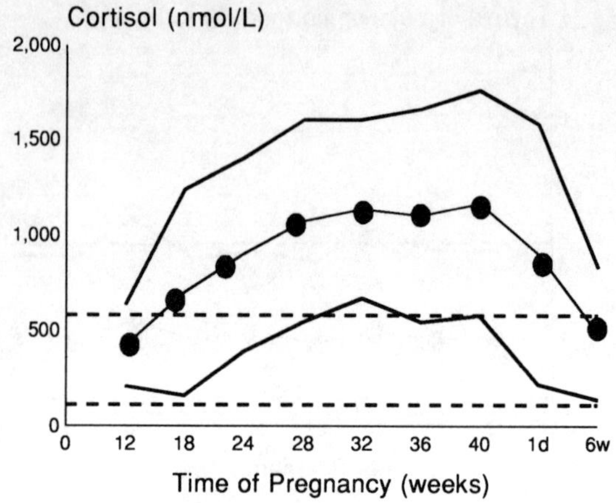

Figure I.L.6. Gestational changes in cortisol concentration.

Table I.L.6. Cortisol (nmol/L)			
Time of Gestation	Reference Range	Mean (sd)	Percentage of Control mean
Non Pregnant Control	110 - 580 #	378 (112)	
12 weeks	205 - 632	418 (107)	111
18 weeks	156 - 1240	698 (271)	185
24 weeks	391 - 1407	899 (254)	238
28 weeks	545 - 1605	1075 (265)	284
32 weeks	667 - 1607	1137 (236)	301
36 weeks	543 - 1663	1103 (280)	292
Term	578 - 1762	1169 (296)	309
1 day pp	215 - 1583	899 (342)	238
6 week pp	136 - 828	482 (173)	128

Non-pregnant control reference range - data from OC users excluded

Part II

Gestational Changes in Hematology

II. A. The Effect of Pregnancy on Hematological Indices

Many of the hematological changes in pregnancy are a consequence of changes in blood volume. Blood volume studies in pregnancy which have been reported in the literature should be interpreted with caution. The reference techniques for the measurement of plasma volume and red cell mass involve the use of radioisotopes whose use is contraindicated in a pregnant woman. Consequently, much of the published data is based on suboptimal techniques, particularly the use of Evan's Blue dye as a measure of plasma volume. The unavailability of optimal tracers, compounded by inadequate mixing, induced by the obstruction of inferior vena cava by the gravid uterus in the supine position[1] should lend caution to the interpretation of published data.

PLASMA VOLUME CHANGES

The plasma volume increases throughout pregnancy. The increase starts at about the 6th week of pregnancy, is minimal in the first trimester, but thereafter reaches a maximum at 34 to 36 weeks gestation.[2,3] A mean plasma volume increase of 45 to 55% above baseline has been reported; in the patient of average weight this amounts to an expansion of 1 to 1.5 L. Plasma volume measurements after 34 weeks gestation are problematical with inadequate tracer mixing due to vena cava obstruction, but it is likely that the plasma volume plateaus, but does not decline, in late pregnancy.[2,4] The increment in plasma volume is higher in second and subsequent pregnancies, probably related to the higher birth weight of second and subsequent infants.[3] Similarly, multiple pregnancy results in even higher plasma volume increments.[3]

The mechanism for the increase in plasma volume is poorly understood but probably relates to hormonal changes, possibly associated with an arteriovenous "shunt" effect within the placenta.[5]

CHANGES IN RED CELL MASS

The changes in the red cell mass in pregnancy have been even less adequately studied than plasma volume changes. It appears, however, that after a brief decline in the first trimester the red cell mass increases steadily in a linear fashion to reach non-pregnant values at 30 weeks gestation and maximum values at term.[4,6] The mean increase in red cell mass from the non-pregnant state amounts to 18% (about 200 mL of red cells) in women not receiving iron and folate supplementation and 30% (about 350 mL of red cells) in iron- and folate-supplemented individuals.[3] This significant expansion in the red cell mass and in other tissues, such as the uterus and the fetus itself, creates a considerable demand for iron. As iron is directed away from bone marrow storage sites to these growing tissues, the serum iron and serum ferritin fall to reach a nadir at 28 to 32 weeks.[7,8] This fall is particularly marked in those not given iron supplementation.[7] Iron absorption increases in late pregnancy to provide the required iron but, even with the iron levels in an average Western diet, a negative iron balance results.[4,9] This imbalance can be corrected by oral iron supplementation; however, active treatment for iron deficiency in pregnancy is rarely required and routine prophylaxis should probably only be given on the basis of abnormal red cell indices and subnormal serum ferritin levels and even then iron is usually required only in the last trimester.[4]

Red cell production is controlled largely by erythropoietin. Erythropoietin, which is produced in the kidney in response to tissue hypoxia, increases during pregnancy after 8 weeks gestation.[10] Recent studies have shown, however, that relative to the hematocrit level there is blunted erythropoietin production and erythropoietic activity in early pregnancy.[11] Erythropoietin levels normalize in the third trimester with increased levels late in pregnancy.[11] The mechanism for the relatively decreased erythropoietin production in early pregnancy is not clear, but increased renal oxygen delivery associated with increased erythrocyte 2,3-DPG levels and increased renal blood flow has been proposed.[11] Late in pregnancy, as a greater proportion of cardiac output is directed to the uterus and placenta, a relatively decreased renal blood flow may cause increased erythropoietin production.[11] These changes may be compounded by alterations in placental lactogen production.[3]

The expansion in both plasma volume and red cell mass is corrected to some degree, at delivery, by the blood loss which occurs at that time. Despite this blood loss, normal women with up to 500 mL of blood loss at delivery may still be hypervolemic in the immediate postpartum period and can still tolerate further hemorrhage.[12]

BLOOD COUNT CHANGES IN PREGNANCY

The consequence of the simultaneous alterations in plasma volume and red cell mass is a progressive fall in total body hematocrit throughout pregnancy. The hematocrit has been reported to reach its nadir at about 30 weeks of gestation.[3]

This physiological drop in hematocrit and hemoglobin has been mistakenly called "the anemia of pregnancy".[13] The decline in hematocrit and hemoglobin occurs from the first trimester onward with minimal changes occurring in the second trimester; this is largely due to dilution by the disproportionate expansion of the plasma volume.[4,14,15] Iron supplementation usually prevents further falls in hemoglobin in the third trimester, when hemoglobin levels usually rise slightly.[14,15] True anemia in pregnancy should be defined on the basis of pregnancy-specific reference ranges but a hemoglobin of <100 g/L is usually pathological.[16]

Red cell indices during pregnancy can be affected significantly by iron deficiency, which will produce reductions in MCV, MCH, and MCHC, or by folate deficiency which causes a raised MCV and MCH. In the typical patient who is depleted in neither iron nor folate, a mild increase in MCV occurs; the highest MCV values usually no higher than 102 fL are seen in the last trimester.[16,17] The cause for this mild macrocytosis is not related to folate deficiency; this is felt to be a normal physiological change.[17]

LEUKOCYTE CHANGES

There is a modest increase in the leukocyte count up to 12×10^9/L during pregnancy which predominantly affects the neutrophil component of the differential.[18,19] Typically there is a doubling of the absolute neutrophil count. This change, however, may not reflect a true increase in granulopoiesis but rather increased demargination of circulating neutrophils.[19] The neutrophil changes may also be due to the effects of increased cortisol levels,[18] possibly upon the maturing bone marrow neutrophil pool. The increase in total leukocyte count has been reported to be apparent by the end of the first trimester and to progress through the second trimester only to decline during the last month of gestation.[18] The absolute neutrophil count rises and falls in keeping with the total leukocyte count; the absolute monocyte count has been reported to rise in the last two trimesters, while the absolute lymphocyte count shows a slight, but statistically insignificant, decline from the first trimester onward.[18] Absolute eosinophil and basophil counts have been reported to show slight, but significant, declines progressively throughout pregnancy.[18]

PLATELET COUNT CHANGES

The platelet count has been reported to be unaffected by pregnancy;[20,21] however, other sequential studies have shown a progressive fall of up to 20% in the platelet count with advancing gestation.[18,22] There is, however, considerable variation in individual platelet counts with pregnancy and group decrements have been attributed to a considerable fall in the platelet counts of individual subjects rather than to a generalized population trend.[22]

Sequential study of platelet indices during pregnancy has shown no significant change in the mean platelet volume but a progressive increase in the platelet distribution width (PDW) has been demonstrated.[21] The PDW is a measure of variation in platelet size, i.e., platelet anisocytosis.

The changes in platelet count in pregnancy have been attributed to blood volume expansion[18] and to increased platelet destruction associated with increased thrombopoiesis during the course of a normal pregnancy.[21,23] Such an increased turnover of platelets would explain an increased PDW. It has been further proposed that normal pregnancy produces a compensated coagulopathy within the placenta, resulting in a decreased platelet lifespan.[21] It is this coagulopathy which is thought to be exacerbated in hypertensive and toxemic states when it results in the pathological thrombocytopenia which occurs in these conditions.[21]

COAGULATION FACTOR CHANGES

The multiple changes in coagulation factors that occur in pregnancy all result in a decreased tendency to peripartum hemorrhage. Blood clotting factors increase with the maximal change occurring in factors VII, VIII, X, XII, and fibrinogen.[24]

The result of all these changes is a mild decrease in clotting times using the standard coagulation screening tests, the prothrombin time (PT), activated partial thromboplastin time (APTT), and thrombin time (TT). The rise in coagulation factors is progressive throughout pregnancy and continues even into the postpartum period.[25]

Regulating factors, such as anti-thrombin III, remain essentially unchanged throughout pregnancy but slightly lower levels than normal are usually found.[25]

Overall fibrinolytic activity as measured by the euglobulin lysis time is reduced from the second trimester onward.[26] Despite this observation using sensitive techniques, low concentrations of FDP and D-dimer can be found in normal pregnancy.[26] Plasminogen levels are increased from the second trimester onward, rising even higher in the third trimester.[26] Fibrinolytic activity rapidly returns to normal within 1 hour of placental separation.[27] For this reason it is postulated that the placenta in some way reduces fibrinolytic activity. Tissue plasminogen activator is reduced in pregnancy and rapidly normalizes after delivery.[26] Tissue plasminogen activator inhibitor (TPAi) increases in the second and third trimesters and it is proposed that the various components of TPAi are derived from the trophoblast and from placental vessels.[26]

In summary, therefore, coagulation activity increases while fibrinolytic activity decreases but fibrinolytic capacity is restored postpartum, thereby diminishing but not abolishing the risk of postpartum venous thrombosis.

THE USE OF HEMATOLOGICAL TESTS IN PREGNANCY

A complete blood count including the hemoglobin, red cell indices, leukocyte count including an automated differential count, and the platelet count should be performed in the early stages of pregnancy. This will reveal preexisting hematological disorders of which the attending physician should be aware. Such disorders would include iron deficiency, thalassemic syndromes, thrombocytopenia, and a variety of other hematological conditions. Any required additional investigation should be performed on the basis of the baseline blood count such that any problems are resolved early in the course of the pregnancy. Additional investigations could include the serum iron and iron binding capacity, serum ferritin, serum folate, and red cell folate and hemoglobin electrophoretic studies.

If the patient is thrombocytopenic, platelet antibody studies, coagulation tests, and even a bone marrow aspiration or biopsy may be required.

For a patient with significant leukopenia or leukocytosis in the absence of infection, close scrutiny of the blood smear should be performed to detect immature leukocyte forms. If immature leukocyte forms are detected early in pregnancy, the patient should be referred for a hematological opinion. The presence of the occasional metamyelocyte in the blood smear of an otherwise healthy patient later in pregnancy should not be a cause for alarm. In patients with unexplained leukopenia or leukocytosis with no other hematological abnormalities and an absence of immature leukocyte forms, a follow-up blood count within a few days is advised. A hematological opinion should be sought if the abnormality persists.

In the uncomplicated pregnancy, the blood count should be repeated at 28 to 30 weeks when the blood count will be at its nadir. If the patient is not receiving iron supplementation, the red cell indices, particularly the MCV and MCH, should be closely scrutinized and compared with reference values. Any suggestion of red cell microcytosis or hypochromia which did not exist at the start of pregnancy should be a signal for the medical attendant to commence oral iron supplementation. Similarly, abnormal red cell macrocytosis should be investigated and, if appropriate, folate supplementation added. Any other new hematological findings at this stage such as leukopenia, leukocytosis, or thrombocytopenia should be investigated as previously outlined. Persistent and progressive thrombocytopenia should be investigated by a hematologist.

Coagulation studies are not part of the routine investigation and follow-up of a pregnant woman. The PT, APTT, fibrinogen quantitation, thrombin time, and D-dimer quantitation are usually only required when some unexpected complication of pregnancy occurs. Such conditions could include toxemia, HELLP syndrome, placental abruption, or fetal death *in utero*. In these severe complications, a severe consumptive coagulopathy (DIC) can ensue and sequential measurement of these coagulation screening procedures together with a platelet count can give very useful indications of clinical severity and clinical trends. It is important to have normal reference values for coagulation tests in the full-term pregnant patient, most particularly for fibrinogen quantitation. Fibrinogen quantitation is normally performed as part of a DIC screen and the inadvertent use of normal adult fibrinogen levels may impair the recognition of early DIC. Unexplained abnormalities in the coagulation screen should be interpreted with the help of a hematologist.

NOTE FOR HEMATOLOGY TESTS

The ranges presented for pregnancy are derived from the whole study group and are not separated in iron supplemented and non-supplemented. PT and APTT are both reagent and instrumentation dependent. It is advised that comparative use of these normal range tables be derived from a comparison of the mean of the non-pregnant range shown in this study with the results of similar tests in one's own laboratory. The ''percentage of control mean'' column can then be used to establish comparative values for other institutions at various stages of pregnancy. *We emphasize that the data shown here must not be transposed directly to the reader's own practice.*

REFERENCES

1. **Chesley, L. C. and Duffus, G. M.,** Posture and apparent plasma volume in late pregnancy, *J. Obstet. Gynecol.*, 78, 406, 1971.
2. **Pirani, B. B. K. and Campbell, D. M.,** Plasma volume in normal first pregnancy, *J. Obstet. Gynaecol. Br. Commonw.*, 80, 884, 1973.
3. **Hytten, F.,** Blood volume changes in normal pregnancy, *Clin. Haematol.*, 14, 601, 1985.
4. **Bentley, D. P.,** Iron metabolism and anaemia in pregnancy, *Clin. Haematol.*, 14, 613, 1985.
5. **Longo, L. D.,** Maternal blood volume and cardiac output during pregnancy: a hypothesis of endocrinologic control, *Am. J. Physiol.*, 245, 720, 1983.
6. **Berlin, M. I., Goetsch, C., Hyde, G. M., and Parsons, R. J.,** The blood volume in pregnancy as determined by ^{32}P labelled red blood cells, *Surg. Gynecol. Obstet.*, 97, 173, 1953.
7. **Fenton, V., Cavill, I., and Fisher, J.,** Iron stores in pregnancy, *Br. J. Haematol.*, 37, 145, 1977.
8. **Foulkes, J. and Goldie, D. J.,** The use of ferritin to assess the need for iron supplementation in pregnancy, *J. Obstet. Gynaecol.*, 3, 11, 1982.
9. **Heinrich, M. C.,** Intestinal iron absorption in man — methods of measurement, dose relationship, diagnostic and therapeutic applications, in *Iron Deficiency; Pathogenesis, Clinical Aspects Therapy,* Hallberg, L., Harwerth, H. G., and Vanotti, A., Eds., Academic Press, London, 1970, 213.
10. **Cotes, P. M. and Canning, C. E.,** Changes in serum immunoreactive erythropoietin during the menstrual cycle and normal pregnancy, *Br. J. Obstet. Gynaecol.*, 90, 304, 1983.

11. **Beguin, Y., Lipscel, G., Thoumsin, H., and Fillet, G.,** Blunted erythropoietin production and decreased erythropoiesis in early pregnancy, *Blood,* 78, 89, 1991.
12. **Pritchard, J. A.,** Changes in the blood volume during pregnancy and delivery, *Anesthesiology,* 26, 393, 1965.
13. **Peck, T. M. and Arias, F.,** Hematologic changes associated with pregnancy, *Clin. Obstet. Gynecol.,* 22, 785, 1979.
14. **De Leeuw, N. K. M., Lowenstein, L., and Hsieh, Y.-S.,** Iron deficiency and hydremia in normal pregnancy, *Medicine,* 45, 291, 1966.
15. **Puolakka, J., Janne, O., Pakarinen, A., Jarvinen, P. A., and Vihko, R.,** Serum ferritin as a measure of iron stores during and after normal pregnancy with and without iron supplements, *Acta Obstet. Gynecol. Scand. Suppl.,* 95, 43, 1980.
16. **De Leeuw, N. K. M. and Brunton, L.,** Maternal hematologic changes, iron metabolism, and anemias in pregnancy, in *Perinatal Medicine: The Basic Science Underlying Clinical Practice,* Goodwin, J. W., Godden, J. O., and Chance, G. W., Eds., Williams & Wilkins, Baltimore, 1976, 425.
17. **Chanarin, I., McFadyen, I. R., and Kyle, R.,** The physiological macrocytosis of pregnancy, *Br. J. Obstet. Gynaecol.,* 84, 504, 1977.
18. **Pitkin, R. M. and Witte, D. L.,** Platelet and leukocyte counts in pregnancy, *JAMA,* 242, 2696, 1979.
19. **Hiss, R. G.,** *Blood Disorders in Pregnancy,* Laros, R. K., Ed., Lea & Febiger, Philadelphia, 1986.
20. **Fenton, V., Saunders, K., and Cavill, I.,** The platelet count in pregnancy, *J. Clin. Pathol.,* 30, 68, 1977.
21. **Tygart, S. G., McRoyan, D. K., Spinnato, J. A., McRoyan, C. J., and Kitay, D. Z.,** Longitudinal study of platelet indices during normal pregnancy, *Am. J. Obstet. Gynecol.,* 154, 883, 1986.
22. **Sill, P. R. and Walker, W.,** Platelet values during normal pregnancy, *Br. J. Obstet. Gynaecol.,* 92, 480, 1985.
23. **Fay, R. A., Hughes, A. O., and Farran, N. T.,** Platelets in pregnancy: hyperdestruction in pregnancy, *Obstet. Gynecol.,* 61, 238, 1983.
24. **Bonnar, J.,** Hemostasis and coagulation in pregnancy, in *Haemostasis and Thrombosis,* Bloom, A. L. and Thomas, D. P., Eds., Churchill Livingstone, Edinburgh, 1981.
25. **Stirling, Y., Woolf, L., North, W. R. S., Seghatchian, M. J., and Meade, T. W.,** Haemostasis in normal pregnancy, *Thromb. Haemostas.,* 52, 176, 1984.
26. **Wright, J. G., Cooper, P., Astedt, B., Lecander, I., Wilde, J. T., Preston, F. E., and Greaves, M.,** Fibrinolysis during normal human pregnancy: complex inter-relationships between plasma levels of tissue plasminogen activator and inhibitors and the euglobulin clot lysis time, *Br. J. Haematol.,* 69, 253, 1988.
27. **Bonnar, J., McNicol, G. P., and Douglas, A. S.,** Coagulation and fibrinolytic mechanisms during and after normal childbirth, *Br. Med. J.,* ii, 200, 1970.

II.A.1. HEMOGLOBIN

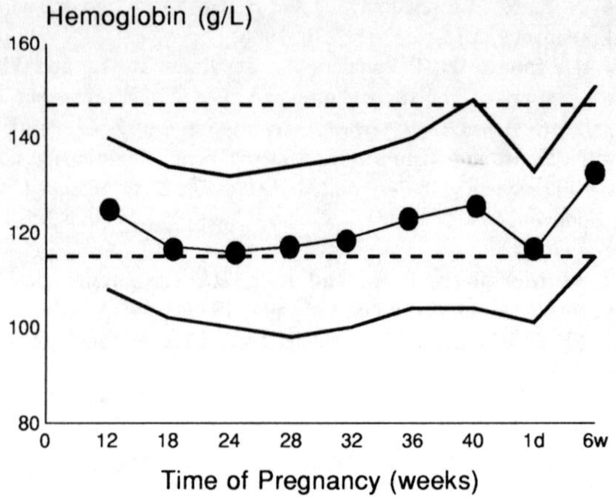

Figure II.A.1. Gestational changes in hemoglobin concentration.

Table II.A.1. Hemoglobin (g/L)			
Time of Gestation	Reference Range	Mean (sd)	Percentage of Control mean
Non Pregnant Control	115 - 147	131 (8)	
12 weeks	108 - 140	124 (8)	95
18 weeks	102 - 134	118 (8)	90
24 weeks	100 - 132	116 (8)	89
28 weeks	98 - 134	116 (9)	89
32 weeks	100 - 136	118 (9)	90
36 weeks	104 - 140	122 (9)	93
Term	104 - 148	126 (11)	96
1 day pp	101 - 133	117 (8)	89
6 week pp	115 - 151	133 (9)	102

II.A.2. ERYTHROCYTE COUNT

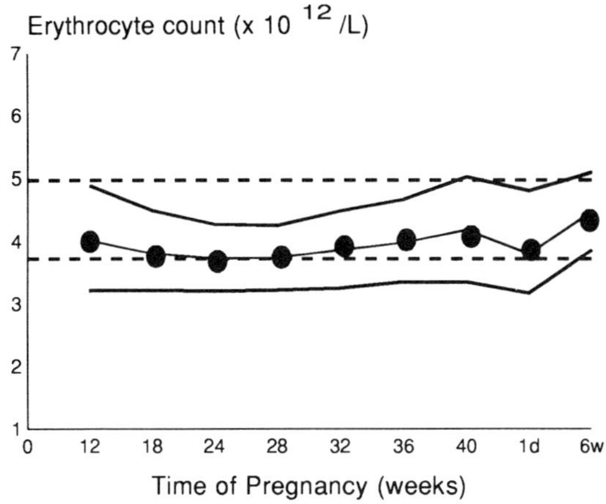

Figure II.A.2. Gestational changes in erythrocyte count.

Table II.A.2. Erythrocyte Count (X 10^{12}/L)			
Time of Gestation	Reference Range	Mean (sd)	Percentage of Control mean
Non Pregnant Control	3.72 - 4.98	4.33 (0.33)	
12 weeks	3.21 - 4.89	4.05 (0.42)	94
18 weeks	3.22 - 4.50	3.86 (0.32)	89
24 weeks	3.20 - 4.28	3.74 (0.27)	86
28 weeks	3.22 - 4.26	3.74 (0.26)	86
32 weeks	3.25 - 4.49	3.87 (0.31)	89
36 weeks	3.35 - 4.67	4.01 (0.33)	93
Term	3.35 - 5.03	4.19 (0.42)	97
1 day pp	3.17 - 4.81	3.99 (0.41)	92
6 week pp	3.86 - 5.10	4.48 (0.31)	103

II.A.3. HEMATOCRIT

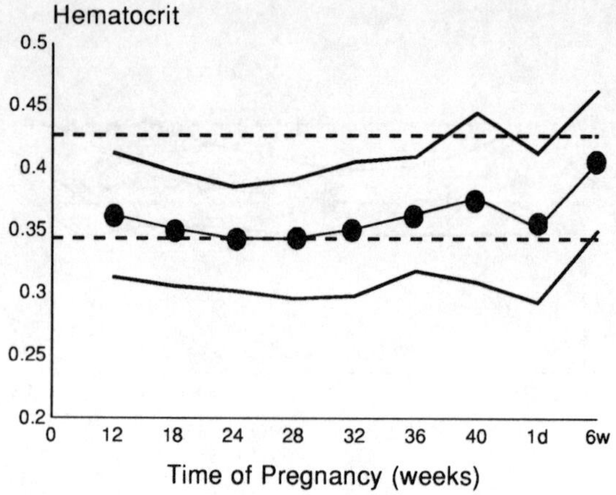

Figure II.A.3. Gestational changes in hematocrit.

Table II.A.3. Hematocrit

Time of Gestation	Reference Range	Mean (sd)	Percentage of Control mean
Non Pregnant Control	0.343 - 0.426	0.387 (0.021)	
12 weeks	0.312 - 0.412	0.362 0.025)	94
18 weeks	0.305 - 0.397	0.351 (0.023)	91
24 weeks	0.301 - 0.385	0.343 0.021)	89
28 weeks	0.295 - 0.391	0.343 (0.024)	89
32 weeks	0.297 - 0.405	0.351 (0.027)	91
36 weeks	0.317 - 0.409	0.363 (0.023)	94
Term	0.308 - 0.444	0.376 (0.034)	97
1 day pp	0.292 - 0.412	0.352 (0.030)	91
6 week pp	0.350 - 0.462	0.406 (0.028)	105

II.A.4. MEAN CORPUSCULAR VOLUME

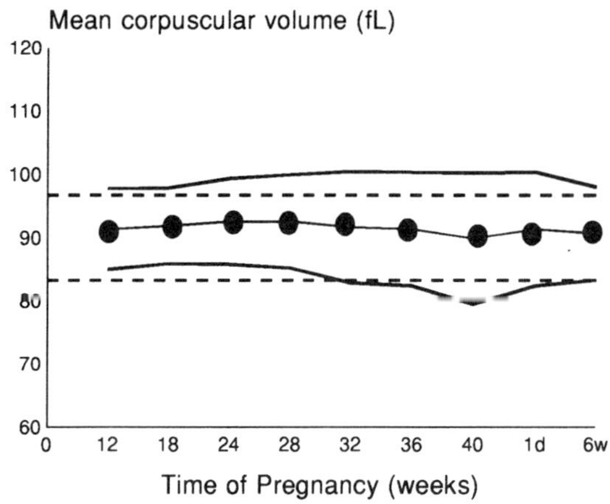

Figure II.A.4. Gestational changes in mean corpuscular volume.

Table II.A.4. Mean Corpuscular Volume (fL)

Time of Gestation	Reference Range	Mean (sd)	Percentage of Control mean
Non Pregnant Control	83.3 - 96.8	90.3 (3.4)	
12 weeks	85.0 - 97.8	91.4 (3.2)	101
18 weeks	85.9 - 97.9	91.9 (3.0)	102
24 weeks	85.8 - 99.4	92.6 (3.4)	103
28 weeks	85.2 - 100.0	92.6 (3.7)	103
32 weeks	82.9 - 100.5	91.7 (4.4)	102
36 weeks	82.4 - 100.4	91.4 (4.5)	101
Term	79.5 - 100.3	89.9(5.2)	100
1 day pp	82.4 - 100.4	91.4 (4.5)	101
6 week pp	83.3 - 98.1	90.7 (3.7)	100

II.A.5. MEAN CORPUSCULAR HEMOGLOBIN

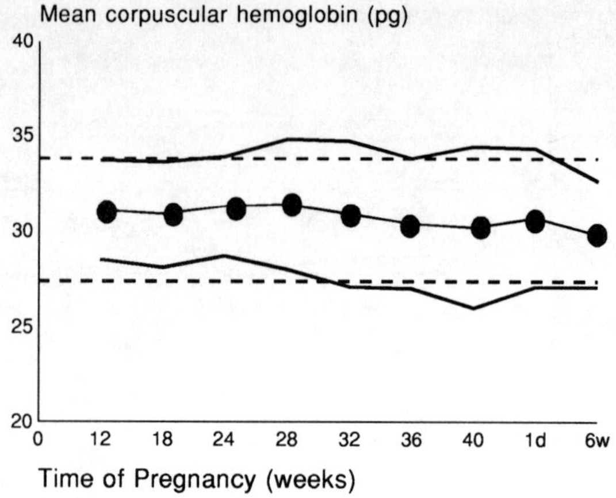

Figure II.A.5. Gestational changes in mean corpuscular hemoglobin.

Table II.A.5. Mean Corpuscular Hemoglobin (pg)			
Time of Gestation	Reference Range	Mean (sd)	Percentage of Control mean
Non Pregnant Control	27.4 - 33.8	30.6 (1.6)	
12 weeks	28.5 - 33.7	31.1 (1.3)	102
18 weeks	28.1 - 33.7	30.9 (1.4)	101
24 weeks	28.7 - 33.9	31.3 (1.3)	102
28 weeks	28.0 - 34.8	31.4 (1.7)	103
32 weeks	27.1 - 34.7	30.9 (1.9)	101
36 weeks	27.0 - 33.8	30.4 (1.7)	99
Term	26.0 - 34.4	30.2 (2.1)	99
1 day pp	27.1 - 34.3	30.7 (1.8)	100
6 week pp	27.0 - 32.6	29.8 (1.4)	97

II.A.6. MEAN CORPUSCULAR HEMOGLOBIN CONCENTRATION

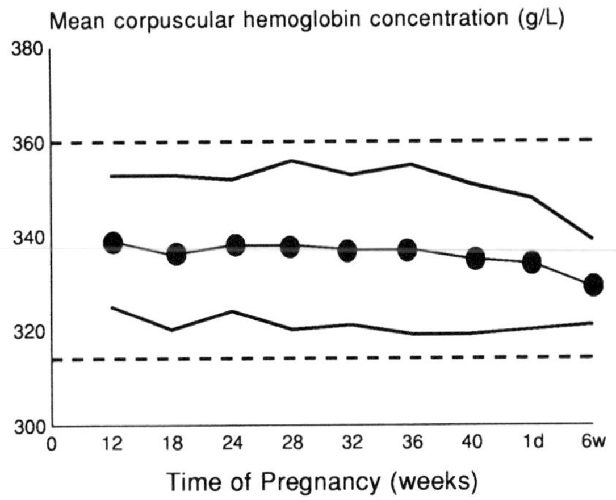

Figure II.A.6. Gestational changes in mean corpuscular hemoglobin concentration.

Table II.A.6. Mean Corpuscular Hemoglobin Concentration (g/L)			
Time of Gestation	Reference Range	Mean (sd)	Percentage of Control mean
Non Pregnant Control	314 - 360	339 (9)	
12 weeks	325 - 353	339 (7)	100
18 weeks	320 - 353	336 (8)	99
24 weeks	324 - 352	338 (7)	100
28 weeks	320 - 356	338 (9)	100
32 weeks	321 - 353	337 (8)	99
36 weeks	319 - 355	337 (9)	99
Term	319 - 351	335 (8)	99
1 day pp	320 - 348	334 (7)	99
6 week pp	321 - 339	329 (4)	97

II.A.7. RED CELL DISTRIBUTION WIDTH

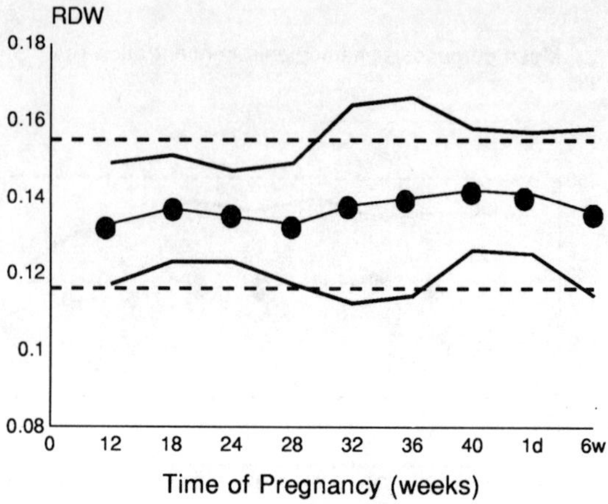

Figure II.A.7. Gestational changes in red cell distribution width (RDW).

Table II.A.7. RDW			
Time of Gestation	Reference Range	Mean (sd)	Percentage of Control mean
Non Pregnant Control	0.116 - 0.155	0.129 (0.008)	
12 weeks	0.117 - 0.149	0.133 (0.008)	103
18 weeks	0.123 - 0.151	0.137 (0.007)	106
24 weeks	0.123 - 0.147	0.135 (0.006)	105
28 weeks	0.117 - 0.149	0.133 (0.008)	103
32 weeks	0.112 - 0.164	0.138 (0.013)	107
36 weeks	0.114 - 0.166	0.14 (0.013)	109
Term	0.126 - 0.158	0.142 (0.008)	110
1 day pp	0.125 - 0.157	0.141 (0.008)	109
6 week pp	0.114 - 0.158	0.136 (0.011)	105

II.A.8. LEUKOCYTE COUNT

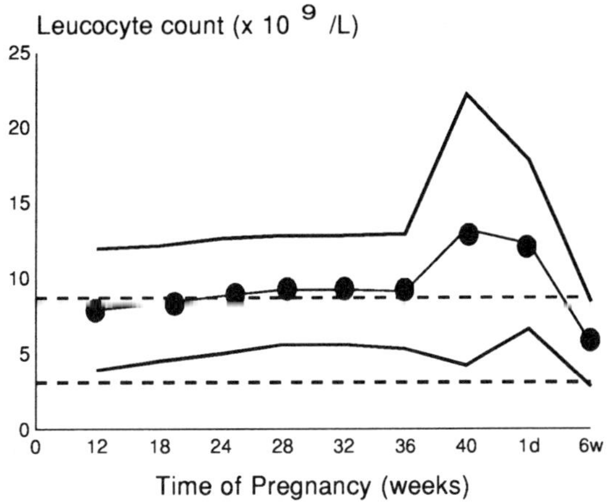

Figure II.A.8. Gestational changes in leucocyte count.

Table II.A.8. Leucocyte Count (X 10^9/L)			
Time of Gestation	Reference Range	Mean (sd)	Percentage of Control mean
Non Pregnant Control	3.1 - 8.7	5.5 (1.3)	
12 weeks	3.9 - 11.9	7.9 (2.0)	144
18 weeks	4.5 - 12.1	8.3 (1.9)	151
24 weeks	5.0 - 12.6	8.8 (1.9)	160
28 weeks	5.6 - 12.8	9.2 (1.8)	167
32 weeks	5.6 - 12.8	9.2 (1.8)	167
36 weeks	5.3 - 12.9	9.1 (1.9)	165
Term	4.2 - 22.2	13.2 (4.5)	240
1 day pp	6.6 - 17.8	12.2 (2.8)	222
6 week pp	2.8 - 8.4	5.6 (1.4)	102

II.A.9. LYMPHOCYTE COUNT (AUTOMATED)

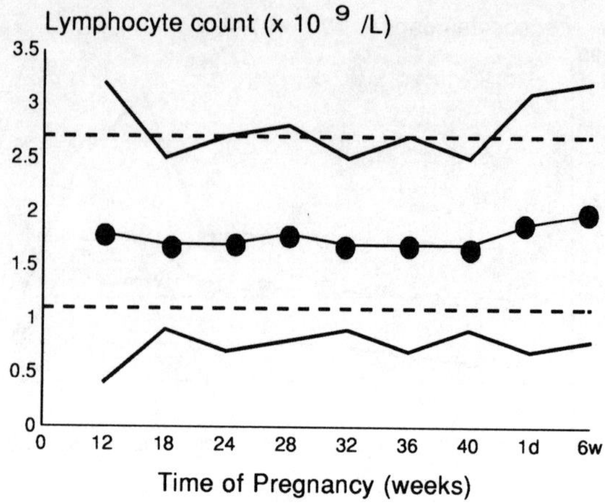

Figure II.A.9. Gestational changes in auto lymphocyte count.

Table II.A.9. Lymphocyte Count (X 10^9/L)			
Time of Gestation	Reference Range	Mean (sd)	Percentage of Control mean
Non Pregnant Control	1.1 - 2.7	1.8 (0.4)	
12 weeks	0.4 - 3.2	1.8 (0.7)	100
18 weeks	0.9 - 2.5	1.7 (0.4)	94
24 weeks	0.7 - 2.7	1.7 (0.5)	94
28 weeks	0.8 - 2.8	1.8 (0.5)	100
32 weeks	0.9 - 2.5	1.7 (0.4)	94
36 weeks	0.7 - 2.7	1.7 (0.5)	94
Term	0.9 - 2.5	1.7 (0.4)	94
1 day pp	0.7 - 3.1	1.9 (0.6)	106
6 week pp	0.8 - 3.2	2.0 (0.6)	111

II.A.10. MONOCYTE COUNT (AUTOMATED)

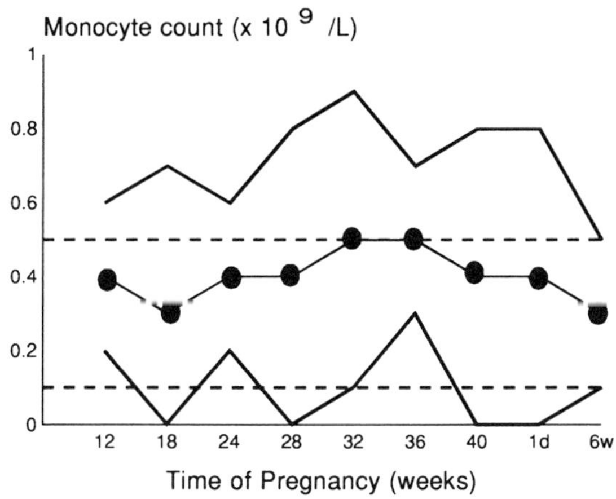

Figure II.A.10. Gestational changes in auto monocyte count.

Table II.A.10. Monocyte Count (X 10^9/L)			
Time of Gestation	Reference Range	Mean (sd)	Percentage of Control mean
Non Pregnant Control	0.1 - 0.5	0.2 (0.1)	
12 weeks	0.2 - 0.6	0.4 (0.1)	200
18 weeks	0 - 0.7	0.3 (0.2)	150
24 weeks	0.2 - 0.6	0.4 (0.1)	200
28 weeks	0 - 0.8	0.4 (0.2)	200
32 weeks	0.1 - 0.9	0.5 (0.2)	250
36 weeks	0.3 - 0.7	0.5 (0.1)	250
Term	0 - 0.8	0.4 (0.2)	200
1 day pp	0 - 0.8	0.4 (0.2)	200
6 week pp	0.1 - 0.5	0.3 (0.1)	150

II.A.11. GRANULOCYTE COUNT (AUTOMATED)

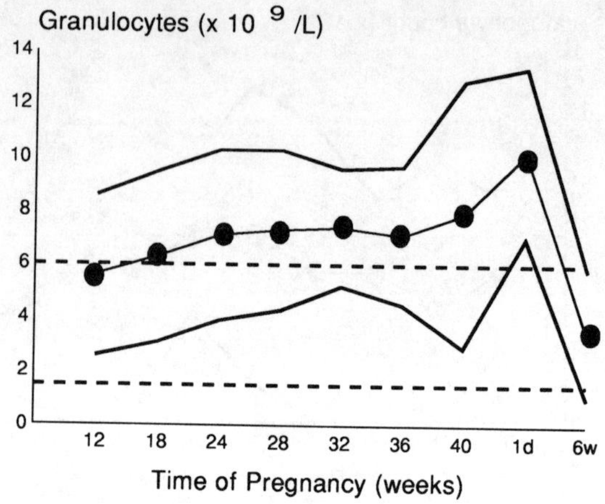

Figure II.A.11. Gestational changes in auto granulocyte count.

Table II.A.11. Granulocyte Count (X 10^9/L)			
Time of Gestation	Reference Range	Mean (sd)	Percentage of Control mean
Non Pregnant Control	1.5 - 6.0	3.3 (1.0)	
12 weeks	2.6 - 8.6	5.6 (1.5)	170
18 weeks	3.1 - 9.5	6.3 (1.6)	191
24 weeks	3.9 - 10.3	7.1 (1.6)	215
28 weeks	4.3 - 10.3	7.3 (1.5)	221
32 weeks	5.2 - 9.6	7.4 (1.1)	224
36 weeks	4.5 - 9.7	7.1 (1.3)	215
Term	2.9 - 12.9	7.9 (2.5)	239
1 day pp	7.0 - 13.4	10.2 (1.6)	309
6 week pp	1.0 - 5.8	3.4 (1.2)	103

II.A.12. NEUTROPHIL COUNT (MANUAL)

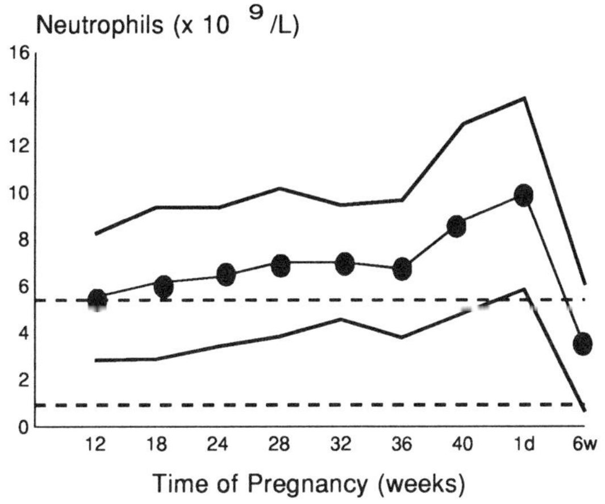

Figure II.A.12. Gestational changes in manual neutrophil count.

Table II.A.12. Neutrophil Count (X 10^9/L)			
Time of Gestation	Reference Range	Mean (sd)	Percentage of Control mean
Non Pregnant Control	0.92 - 5.40	3.16 (1.12)	
12 weeks	2.82 - 8.23	5.52 (1.35)	175
18 weeks	2.87 - 9.39	6.13 (1.63)	194
24 weeks	3.40 - 9.36	6.38 (1.49)	202
28 weeks	3.82 - 10.18	7.00 (1.59)	222
32 weeks	4.55 - 9.47	7.01 (1.23)	222
36 weeks	3.79 - 9.67	6.73 (1.47)	213
Term	4.81 - 12.93	8.81 (2.03)	279
1 day pp	5.84 - 14.00	9.92 (2.04)	314
6 week pp	0.62 - 6.04	3.32 (1.36)	105

II.A.13. PLATELET COUNT

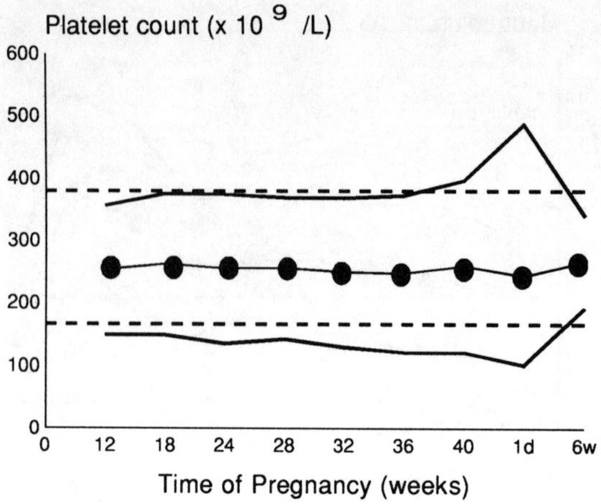

Figure II.A.13. Gestational changes in platelet count.

Table II.A.13. Platelet Count (X 10^9/L)			
Time of Gestation	Reference Range	Mean (sd)	Percentage of Control mean
Non Pregnant Control	166 - 381	259 (54)	
12 weeks	149 - 357	253 (52)	98
18 weeks	148 - 376	262 (57)	101
24 weeks	135 - 375	255 (60)	98
28 weeks	142 - 370	256 (57)	99
32 weeks	129 - 369	249 (60)	96
36 weeks	121 - 373	247 (63)	95
Term	121 - 397	259 (69)	100
1 day pp	101 - 485	243 (71)	94
6 week pp	192 - 340	266 (37)	103

II.A.14. MEAN PLATELET VOLUME

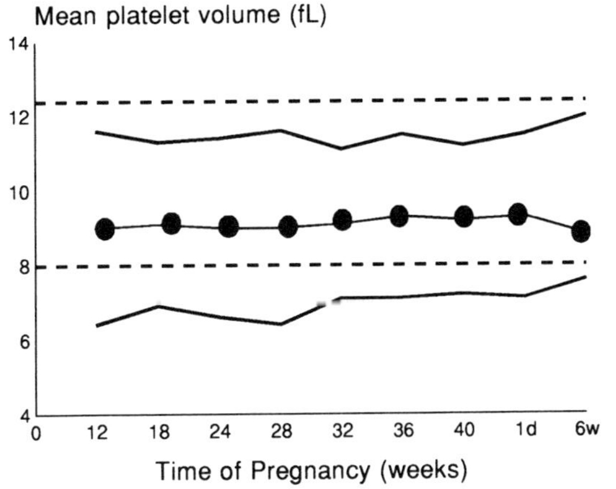

Figure II.A.14. Gestational changes in mean platelet volume.

Table II.A.14. Mean Platelet Volume (fL)			
Time of Gestation	Reference Range	Mean (sd)	Percentage of Control mean
Non Pregnant Control	8.0 - 12.4	10.4 (1.1)	
12 weeks	6.4 - 11.6	9.0 (1.3)	87
18 weeks	6.9 - 11.3	9.1 (1.1)	88
24 weeks	6.6 - 11.4	9.0 (1.2)	87
28 weeks	6.4 - 11.6	9.0 (1.3)	87
32 weeks	7.1 - 11.1	9.1 (1.0)	88
36 weeks	7.1 - 11.5	9.3 (1.1)	89
Term	7.2 - 11.2	9.2 (1.0)	88
1 day pp	7.1 - 11.5	9.3 (1.1)	89
6 week pp	7.6 - 12.0	8.8 (1.1)	85

II.A.15. PROTHROMBIN TIME

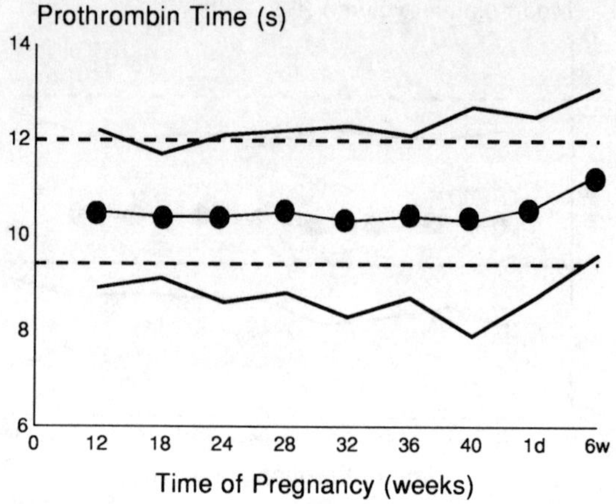

Figure II.A.15. Gestational changes in prothrombin time.

Table II.A.15. Prothrombin Time (s)			
Time of Gestation	Reference Range	Mean (sd)	Percentage of Control mean
Non Pregnant Control	9.4 - 12.0	10.6 (0.6)	
12 weeks	8.9 - 12.2	10.5 (0.8)	99
18 weeks	9.1 - 11.7	10.4 (0.6)	98
24 weeks	8.6 - 12.1	10.4 (0.9)	98
28 weeks	8.8 - 12.2	10.5 (0.8)	99
32 weeks	8.3 - 12.3	10.3 (1.0)	97
36 weeks	8.7 - 12.1	10.4 (0.8)	98
Term	7.9 - 12.7	10.3 (1.2)	97
1 day pp	8.7 - 12.5	10.6 (1.0)	100
6 week pp	9.6 - 13.1	11.3 (0.9)	107

II.A.16. ACTIVATED PARTIAL THROMBOPLASTIN TIME

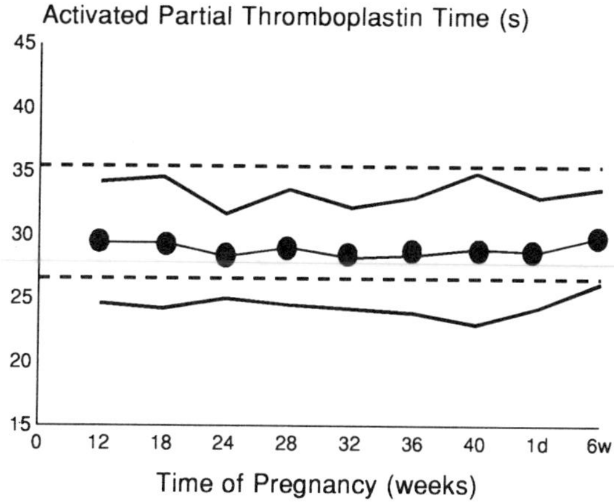

Figure II.A.16. Gestational changes in activated partial thromboplastin time

Table II.A.16. Activated Partial Thromboplastin Time (s)

Time of Gestation	Reference Range	Mean (sd)	Percentage of Control mean
Non Pregnant Control	26.6 - 35.5	31.0 (2.3)	
12 weeks	24.6 - 34.2	29.4 (2.4)	95
18 weeks	24.2 - 34.6	29.4 (2.6)	95
24 weeks	25.0 - 31.7	28.3 (1.7)	91
28 weeks	24.5 - 33.6	29.0 (2.3)	94
32 weeks	24.2 - 32.2	28.2 (2.0)	91
36 weeks	23.9 - 33.0	28.4 (2.3)	92
Term	23.0 - 34.9	28.9 (3.0)	93
1 day pp	24.3 - 33.0	28.7 (2.2)	92
6 week pp	26.2 - 33.7	29.9 (1.9)	97

II.A.17. FIBRINOGEN

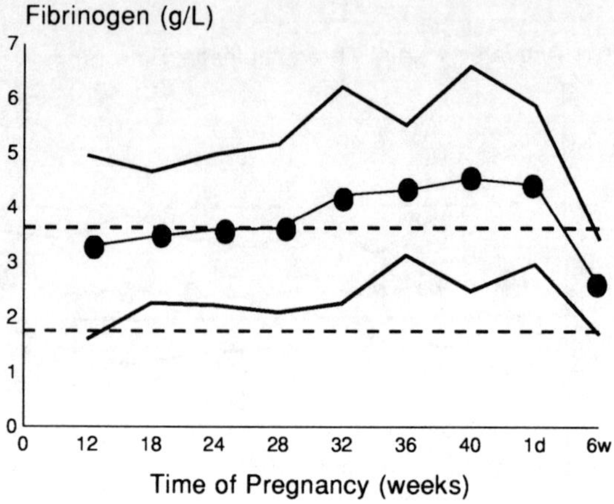

Figure II.A.17. Gestational changes in fibrinogen concentration.

Table II.A.17. Fibrinogen (g/L)

Time of Gestation	Reference Range	Mean (sd)	Percentage of Control mean
Non Pregnant Control	1.76 - 3.64	2.75 (0.51)	
12 weeks	1.60 - 4.96	3.28 (0.84)	119
18 weeks	2.26 - 4.66	3.46 (0.60)	126
24 weeks	2.22 - 4.98	3.60 (0.69)	131
28 weeks	2.09 - 5.17	3.63 (0.77)	132
32 weeks	2.26 - 6.22	4.24 (0.99)	154
36 weeks	3.13 - 5.53	4.33 (0.60)	157
Term	2.48 - 6.60	4.54 (1.03)	165
1 day pp	2.97 - 5.89	4.43 (0.73)	161
6 week pp	1.7 - 3.42	2.56 (0.43)	93

Part III

Reference Ranges for Non-Pregnant Adult Women

III.A. The Non-Pregnant Control Study

RECRUITMENT OF CONTROL SUBJECTS

We recruited 121 healthy non-pregnant women of reproductive age to establish non-pregnant control reference ranges. Our control subjects were of similar socioeconomic status to our pregnant subjects. They were recruited by poster advertisements in the hospital, and were predominantly professional and technical hospital and university personnel, who would be expected to be reasonably knowledgeable about health and nutrition. The only exclusion criteria were acute or chronic illness, alcoholism, heavy smoking, or drug use. Information on oral contraceptive use, smoking habits, and exercise or alcohol ingestion within 24 hours of study was obtained by questionnaire. A copy of the test results was sent to each subject on completion of the study and to her physician if an abnormal result required follow-up. No other incentive was offered.

A single, morning, fasting blood specimen was obtained from each subject with the standardized collection protocol used throughout the pregnancy study (Section IV.B.).

CONTROL OF FACTORS WHICH CAN MODIFY PLASMA CONCENTRATIONS OF ANALYTES

Preanalytic factors[1-3] which may affect the establishment of reference ranges may be considered in two categories; specimen collection procedures could be controlled by a standardized protocol and individual subject lifestyle and activities which are variable.

Procedural and technical variations in specimen collection include the effect of fasting, time of collection, posture and hemostasis during collection, mode and sequence of collection, collection devices, specimen containers, and anticoagulants. All of these factors were carefully controlled in our study by strict adherence to a standardized specimen collection protocol.

Individual lifestyle differences include use of oral contraception, smoking, recent exercise, recent alcohol use, and use of nutritional supplements. These were studied, but not controlled, in our study.

CHARACTERISTICS OF STUDY SAMPLE

The control group included 121 women aged 20 to 49 years. The mean (sd) age was 33 (7) years; 30 of the women were using oral contraceptives. Ten were smokers, none more than a pack per day. Eighteen had ingested an alcoholic beverage and 60 had exercised within 24 hours of specimen collection. The distribution among the categories is shown in Tables III.B.1. and III.B.2.

AGE

The mean age was lower in the oral contraceptive users (29 years) than non-contraceptive users (35 years) and in smokers (26 years) than non-smokers (34 years). There was no significant difference in age between those who had or had not ingested alcohol or exercised prior to the study.

ORAL CONTRACEPTIVE USE

Thirty women were using oral contraceptives. The majority were using a form of phasic low-dose oral contraceptive. Since the individuals volunteered at their own convenience to participate in the study, specimen collection was not restricted to any one phase of the contraceptive cycle.

EXERCISE

Half of our study group exercised within 24 hours before specimen collection. Types of exercise included walking for 1 to 2 hours, running, swimming, aerobics class, training on gym equipment (rowing machine, bicycle, weights), cycling, tennis, squash, and badminton.

ALCOHOL AND SMOKING

The subjects who smoked or had used alcohol were 8.3 and 14.8% of the group. Eighteen subjects had ingested some alcohol within 24 hours of testing. This was in most cases a glass of wine or beer with dinner. Of the ten smokers, three smoked 20/day, three smoked 10/day, and the others less than 5/day.

III.B. Lifestyle Factors that can Modify Test Results

THE EFFECT OF PREANALYTICAL AND LIFESTYLE VARIABLES ON TEST RESULTS

The effect of preanalytical factors on laboratory tests[1-3] and the possible effect of these factors on our control reference data are important considerations. Use of oral contraceptives, cigarettes, and alcohol may independently alter biochemical and hematological parameters, while the interaction of synergistic and antagonistic effects of these factors adds complexity to test interpretation. Although minor differences or trends were found for several variables, in consideration of the many analytes studied, variance between groups or correlation coefficients were only considered important if there was less than a 1 in 1000 probability that the difference was due to chance.

AGE

Although change in concentration of several biochemical indices has been reported to occur in women with increasing age within the age range (20 to 49 years) of our study group, statistically significant correlation with age was shown for few of the analytes which we measured. Slight decreases with age have been reported for serum total protein, albumin, total calcium, and phosphorus,[1] while urea, creatinine, uric acid, total cholesterol, and triglyceride reportedly increase.[1] The 1-hour plasma glucose concentration following a loading dose increases by about 0.4 mmol/L per decade.[1]

Within our control group, glucose was the only variable to show a consistent pattern with increasing age. A positive correlation of fasting glucose with age was found for both the non-oral contraceptive group ($r = 0.35$ $p < 0.001$) and for the whole group ($r = 0.34$ $p < 0.001$).

Although magnesium also showed a positive correlation for the group as a whole ($r = 0.28$ $p < 0.002$), when the non-oral contraceptive users were analyzed alone, no correlation was found. In contrast, ceruloplasmin, which showed no age effect for the group as a whole, showed a positive correlation with age ($r = 0.34$ $p < 0.001$) when the oral contraceptive users were excluded.

No effect of age was noted for total protein, albumin, total calcium, phosphorus, urea, creatinine, uric acid, total cholesterol, or triglyceride in our group of healthy women within the reproductive age range.

ALCOHOL

A single moderate dose of alcohol does not alter laboratory tests but regular ingestion of even moderate amounts may increase plasma triglyceride concentration and activity of plasma enzymes such as γ-glutamyl transferase.[1]

Regular use of alcohol was an exclusion criterion for the control group. We did not demonstrate any significant differences between the women who had not or those who had (approximately 15%) taken alcohol within 24 hours of study.

SMOKING

Smoking may affect several biochemical and hematological indices, depending on the number of cigarettes smoked and degree of inhalation.[1,4-6]

To compensate for reduced oxygen transport capacity due to higher concentrations of carboxyhemoglobin, the blood erythrocyte count is increased in smokers.[1] Blood leukocyte concentration may be increased by 30% compared with non-smokers.[1]

Nicotine stimulates fluid retention. Slight decreases in plasma total protein and albumin and in serum immunoglobulins IgA, IgM and IgG have been reported.[1] Ceruloplasmin is increased in habitual smokers while transferrin is not.[6] Plasma uric acid, urea, and creatinine are also slightly lower in smokers. This decrease in urea and uric acid may be partially due to reduced food intake due to smoking-induced lack of appetite.[1]

Smoking is a major risk factor for coronary and peripheral vascular disease, altering lipid metabolism and coagulation factors.[4] Smoking has been shown to raise total cholesterol, LDL-cholesterol, and triglycerides, while HDL-cholesterol is lowered.[5]

Smoking and alcohol ingestion may act synergistically on blood biochemistry. While no difference was found between women who did or did not smoke, or between women who did or did not drink alcohol, women who were both smokers and drinkers had lower plasma selenium concentrations than those who neither smoked nor drank.[6]

No effect of smoking on our reference ranges was found as our group of smokers was less than 9% of our subjects, and none were heavy smokers.

EXERCISE

While short-term effects of strenuous exercise are clearly demonstrated for several analytes when blood is obtained within 15 min of exercise completion,[1] a sustained effect of regular moderate exercise is less apparent. Slight increases of serum aspartate aminotransferase, lactate dehydrogenase, and creatine kinase activity and decrease in total cholesterol and triglyceride concentration may be noted.[1]

Despite the fact that half of our subjects had exercised within 24 hours of study, no statistically significant difference between the two groups was found for any analyte measured. Creatine kinase was raised in two subjects, to 334 U/L (40 min of aerobic and weight training) and to 559 U/L (an hour of squash). Both women were retested after several days without similar exercise and were within the reference range.

HORMONAL FACTORS

The normal menstrual cycle is accompanied by changes in plasma concentration of several analytes. Plasma cholesterol, total protein, albumin, and calcium are lowest at the time of ovulation,[1] while plasma fibrinogen, phosphate, and iron are lower at menstruation.[1] We did not attempt to assess stage of menstrual cycle in our controls. However, despite the fact that there was no consistency in our control group as to the phase of contraceptive cycle at the time of specimen collection, very marked effects were found for the group of oral contraceptive users as a whole compared to non-oral contraceptive users.

We found significant differences (p <0.001 on t-test) for 17 of the biochemistry variables between our control subjects who were oral contraceptive users and those who were not (Table III.B.3.). For another seven variables, differences were significant at the 0.05 level or less (Table III.B.4.).

EFFECT OF ORAL CONTRACEPTIVES ON BIOCHEMICAL AND HEMATOLOGICAL TESTS

Oral contraceptives significantly affect hepatic synthesis of proteins and secretion of organic anions, lipid and carbohydrate metabolism, and fluid and electrolyte homeostasis. Altered blood biochemistry may include elevated levels of plasma hormone-binding and other transport proteins, increased lipoprotein concentrations, increased tendency to coagulopathy, altered bilirubin and bile salt metabolism, decreased glucose tolerance, and fluid retention.[4,8-21]

Adverse clinical effects may include depression, edema, jaundice, and increased risk of thromboembolism, stroke, myocardial infarction, cholecystitis, and hypertension.[8-10]

There have been many studies of the biochemical effects of oral contraceptive use. Caution must be used when comparing results between studies, since factors such as the chemical formulation and the stage of contraceptive use as well as laboratory methods and ranges may be quite different.

EFFECT OF ESTROGENS AND PROGESTINS ON PLASMA LIPOPROTEINS

Estrogens tend to increase triglyceride and high-density lipoproteins and lower low-density lipoproteins while progestins tend to oppose these effects. It has been shown that women who use oral contraceptives have higher triglycerides, total and LDL-cholesterol, and lower HDL-cholesterol than those who do not.[4]

Oral estrogen administration clearly increases plasma triglyceride levels in both premenopausal[12] and postmenopausal women.[13,14] The increased triglyceride appears to be distributed within LDL, HDL, and VLDL factions.[12] The main effect is probably due to enhanced hepatic secretion of VLDL ApoB.[15]

Although estrogens lower low-density lipoproteins in postmenopausal women,[13,14] they have little effect on healthy premenopausal women with normal or low LDL levels.[15,16]

Estrogens increase all HDL components due to an increase in the HDL_{2b} fraction.[15] In combined oral contraceptive users, the balance between estrogen and progestin determines whether the HDL level is increased, decreased, or unchanged. The more recent formulations with lower dose estrogen and less androgenic progestins produce less effect on low-density lipoproteins and high-density lipoproteins but triglyceride levels are increased because of decreased progestin anti-estrogen effect.[17]

We found a highly significant difference between oral contraceptive users and non-users for triglycerides, less so for total cholesterol, but did not demonstrate differences for LDL-cholesterol or HDL cholesterol. Our results are similar to a recent study of lipids and lipoproteins in oral contraceptive users.[18]

EFFECT OF ESTROGENS AND PROGESTINS ON PLASMA PROTEINS

Estrogens stimulate hepatic synthesis of a variety of proteins including the hormone-binding globulins (thyroxine-binding globulin, sex-hormone-binding globulin, corticosteroid-binding globulin), metal transport proteins (ceruloplasmin, transferrin),[19,20] antiproteases (α-1-antitrypsin), and proteins involved in coagulation and fibrinolysis (antithrombin III, protein C).[21] Some proteins (orosomucoid, haptoglobin) decrease under the effect of estrogen.[20] The stimulatory effect may also be opposed by progestins and the net change in plasma proteins and lipoproteins will depend on the relative dose and type of estrogen and progestin used.[18,20]

We found significant increases in α-1-antitrypsin, ceruloplasmin, C3c, transferrin, thyroxine-binding globulin, prealbumin, and retinol-binding protein, while only albumin was significantly decreased in the oral contraceptive user group.

EFFECT OF ESTROGENS AND PROGESTINS ON OTHER BIOCHEMICAL OR HEMATOLOGICAL VARIABLES

The effects on other analytes are not as marked as those on proteins and lipids. Plasma concentrations of several analytes have been compared between young women who were or were not oral contraceptive users.[9] Plasma glucose concentrations in oral contraceptive users (4.9 mmol/L) were about 3% lower than non-users (5.0 mmol/L)[9] in contrast to our findings of no significant difference between oral contraceptive users (4.6 mmol/L) and non-oral contraceptive users (4.7 mmol/L).

Plasma *total* bilirubin was 25% lower in oral contraceptive users (6.0 mmol/L) than in non-oral contraceptive users (7.9 μmol/L).[9] We did not find a significant difference in *unconjugated* bilirubin with mean values of 9.8 μmol/L in users and 9.5 μmol/L in non-users.

No difference was found between groups for creatinine or for aspartate aminotransferase in our study or that of Knopp.[9] In both studies, slightly lower alkaline phosphatase levels were found in users than non-users but we did not find the difference statistically significant.

We found slightly lower concentrations of magnesium and phosphate in oral contraceptive users. We also found significantly higher levels of selenium in oral contraceptive users as have others.[7] The differences for other analytes, higher copper and iron, higher total but not free thyroxine, higher cortisol and retinol, are largely explained by the increased levels of their binding or transport proteins.

REFERENCES

1. **Young, D. S. and Bermes, E. W.,** Specimen collection and processing; sources of biological variation, in *Textbook of Clinical Chemistry,* Tietz, N. W., Ed., W. B. Saunders, Philadelphia, 1986, 478.
2. **Solberg, H. E. and Gräsbeck, R.,** Reference values, *Adv. Clin. Chem.,* 27, 2, 1989.
3. **Drosdowsky, M. and Sachs, C.,** Factors to be taken into consideration for blood sampling in view of establishing reference values, *Ann. Biol. Clin.,* 38, 251, 1980.
4. **Webber, L. S., Hunter, S. M., Baugh, J. G., Srinavasan, S. R., Sklov, M. C., and Berenson, G. S.,** The interaction of smoking, oral contraceptive use and cardiovascular risk factor variables in children: the Bogalusa Heart Study, *Am. J. Public Health,* 72, 266, 1982.
5. **Halfon, S. T., Green, M. S., and Heiss, G.,** Smoking status and lipid levels in adults of different ethnic origins: the Jerusalem lipid research clinic program, *Int. J. Epidemiol.,* 13, 177, 1984.
6. **Galdston, M., Feldman, J. G., and Levytska, V.,** Antioxidant activity of serum ceruloplasmin and transferrin available iron-binding capacity in smokers and non-smokers, *Am. Rev. Resp. Dis.,* 135, 783, 1987.
7. **Lloyd, B., Lloyd, R. R., and Clayton, B. E.,** Effect of smoking, alcohol and other factors on the selenium status of a healthy population, *J. Epid. Commun. Health.,* 37, 123, 1983.
8. **Lipsett, M. B., Combs, J. W., Catt, K., and Siegel, D. G.,** Problems in contraception, *Ann. Int. Med.,* 74, 251, 1971.
9. **Knopp, R. H., Bergelin, B. A., Wahl, P. W., Walden, C. E., and Chapman, M. B.,** Clinical chemistry alteration in pregnancy and oral contraceptive use, *Obstet. Gynecol.,* 66, 682, 1985.

10. **van Thiel, D. H. and Gavaler, J. S.,** Pregnancy associated steroids and their effects on the liver, *Sem. Liver Dis.*, 7, 1, 1987.
11. **Webber, L. S., Hunter, S. M., Johnson, C. C., Srinavasan, S. R., and Berenson, G. S.,** Smoking, alcohol and oral contraceptives, *Ann. N.Y. Acad. Sci.*, 623, 138, 1991.
12. **Knopp, R. H., Walden, C. E., Wahl, P. W., Hoover, J. J., Warnick, G. R., Albers, J. J., Ogilvie, J. T., and Hazzard, W. R.,** Oral contraceptive and postmenopausal estrogen effects on lipoprotein triglyceride and cholesterol in an adult female population: relationships to estrogen and progestin potency, *J. Clin. Endocrinol. Metab.*, 53, 1123, 1981.
13. **Farish, E., Fletcher, C. D., Hart, D. M., Teo, C., Alazzawi, F., and Howie, C.,** The effects of conjugated equine oestrogens, with and without a cyclical progestogen on lipoproteins, and HDL subfractions in postmenopausal women, *Acta Endocrinologica. (Copenh)*, 113, 123, 1986.
14. **Sherwin, B. B. and Gelfand, M. M.,** A prospective one-year study of estrogen and progestin in postmenopausal women: effects on clinical symptoms and lipoprotein lipids, *Obstet. Gynecol.*, 73, 759, 1989.
15. **Schaefer, E. J., Foster, D. M., Zech, L. A., Lindgren, F. T., Brewer, H. B., and Levy, R. I.,** The effects of estrogen administration on plasma lipoprotein metabolism in premenopausal females, *J. Clin. Endocrinol. Metab.*, 57, 262, 1983.
16. **Tikkanen, M. J., Nikkila, E. A., Juusi, T., and Spipinen, S.,** Effects of oestradiol and levonorgestrel on lipoprotein lipids and postheparin plasma lipase activities in normolipoproteinaemic women, *Acta Endocrinol.*, 99, 630, 1982.
17. **Crook, D. and Seed, M.,** Endocrine control of plasma lipoprotein metabolism: effect of gonadal steroids, *Bailliere's Clin. Endo. Metab.*, 4, 851, 1988.
18. **Godsland, I. F., Crook, D., Simpson, R., Proudler, T., Felton, C., Lees, B., Anyaoku, V., Devenport, M., and Wynn, V.,** The effects of different formulations of oral contraceptive agents on lipid and carbohydrate metabolism, *N. Engl. J. Med.*, 323, 1375, 1990.
19. **Song, S., Chen, J.-K., He, M.-L., and Fotherby, K.,** Effect of some oral contraceptives on serum concentrations of sex hormone binding globulin and ceruloplasmin, *Contraception*, 39, 385, 1989.
20. **Sobbrio, G. A., Granata, A., Granese, D., D'Arrigo, F., Panacea, A., Nicita, R., Pulle, C., and Trimarchi, F.,** Sex hormone binding globulin, cortisol binding globulin, thyroxine binding globulin, ceruloplasmin: changes in treatment with two oral contraceptives low in oestrogen, *Clin. Exp. Obstet. Gynecol.*, 18, 43, 1991.
21. **Beller, F. K. and Ebert, C.,** Effects of oral contraceptives on blood coagulation, *Obstet. Gynecol. Surv.*, 40, 425, 1985.

III.B.1. USE OF ORAL CONTRACEPTIVES IN CONTROL SUBJECTS

Table III.B.1. Number of subjects using or not oral contraceptives who smoke, or who exercised or drank alcohol within 24 hours of study.

Use of Oral Contraceptives		Exercise #		Alcohol		Smoking	
	Age	None	Some	None	Some	None	Some
No OC (n=91)	35 y	47	44	82	9	83	8
OC (n=30)	29 y	13	16	21	9	28	2
Total		60	60	103	18	111	10

Exercise data missing for one subject

III.B.2. USE OF ALCOHOL IN CONTROL SUBJECTS

Table III.B.2. Number of control subjects in each category grouped by recent exercise or alcohol ingestion, and smoking habits

		Smoking		Exercise #	
		None (n=111)	Some (n=10)	None (n=60)	Some (n=60)
Alcohol	Age	34 y	26 y	34 y	33 y
None (n=103)	33 y	95	8	53	50
Some (n= 18)	31 y	16	2	7	10

Exercise data missing for one subject

III.B.3. EFFECT OF ORAL CONTRACEPTIVE USE ON MEAN VALUES FOR BIOCHEMISTRY ANALYTES ($p < 0.001$)

Table III.B.3. Effect of oral contraceptive use on mean and ± 2 sd range for biochemistry analytes where group differences by Student's *t* test were significant at p <0.0001 for each test.

Analyte	No OC Use (N=91)		OC Use (N=30)	
	Mean (sd)	± 2 sd range	Mean (sd)	± 2 sd range
albumin	42 (2)	38 - 46	40 (2)	36 - 44
α-1-antitrypsin	2.15 (0.42)	1.31 - 2.99	2.91 (0.65)	1.61 - 4.21
ceruloplasmin	252 (38)	176 - 328	414 (59)	296 - 532
complement C3c	0.71 (0.12)	0.47 - 0.95	0.82 (0.18)	0.46 - 1.18
copper	15.4 (2.0)	11.4 - 19.4	26.2 (4.5)	17.2 - 35.2
cortisol	378 (112)	154 - 602	786 (212)	362 - 1210
iron	15 (7)	1 - 29	21 (7)	7 - 35
iron-binding capacity	61 (8)	45 - 77	73 (10)	53 - 93
prealbumin	246 (46)	154 - 338	291 (29)	233 - 349
retinol	1.40 (0.34)	0.72 - 2.08	1.99 (0.29)	1.41 - 2.57
retinol-binding protein	38 (8)	22 - 54	52 (7)	38 - 66
thyroxine	104 (17)	70 - 138	132 (15)	102 - 162
thyroxine-BG	22 (3)	16 - 28	33 (4)	25 - 41
transferrin	2.85 (0.40)	2.05 - 3.65	3.30 (0.36)	2.58 - 4.02
triglyceride	0.88 (0.35)	0.18 - 1.58	1.24 (0.53)	0.18 - 2.30
triiodothyronine	1.9 (0.3)	1.3 - 2.5	2.3 (0.3)	1.7 - 2.9
selenium	1.71 (0.22)	1.27 - 2.15	1.91 (0.25)	1.41 - 2.41

III.B.4. EFFECT OF ORAL CONTRACEPTIVE USE ON MEAN VALUES FOR BIOCHEMISTRY ANALYTES ($p < 0.05$)

Table III.B.4. Effect of oral contraceptive use on mean and ± 2 sd range for biochemistry analytes where group differences by *t*-test were significant at p <0.05 for each test.

Analyte	No OC Use (N=91)		OC Use (N=30)	
	Mean (sd)	± 2 sd range	Mean (sd)	± 2 sd range
α-tocopherol: cholesterol ratio	5.5 (1)	3.5 - 7.5	4.9 (1)	2.9 - 6.9
cholesterol (total)	4.39 (0.84)	2.71 - 6.07	4.77 (1.13)	2.51 - 7.03
creatine kinase	63 (36)	0 - 135	78 (36)	6 - 150
γ-glutamyl transferase	22 (7)	8 - 36	25 (10)	5 - 45
lactate dehydrogenase	398 (65)	268 - 528	370 (58)	254 - 485
magnesium	0.80 (0.06)	0.68 - 0.92	0.77 (0.05)	0.67 - 0.87
phosphate	1.10 (0.13)	0.84 - 1.38	1.02 (0.15)	0.72 - 1.32

III.C. Reference Ranges for Non-Pregnant Women

The reference ranges following were derived from the 121 control subjects described in Section III.A. who met our stated criteria, healthy well-nourished women who were generally fit and active and did not indulge in excess alcohol or cigarette use. Despite the fact that the results following reflect the biochemical and hematological status of these healthy young women, it is important to emphasize that these reference limits for certain analytes do not necessarily imply optimal levels for health.

For example, subclinical iron deficiency, suggested by ferritin concentrations of <12 μmol/L was present in 17% of this group, while in 10% the ferritin level was <6 μmol/L. The upper limit (97.5%) of the total cholesterol range was 6.87 mmol/L. This is probably higher than desired for prevention of atherosclerosis.[1] Similarly, upper limits for LDL-cholesterol and triglycerides and lower limits for HDL-cholesterol may also fall outside the desired limits.

The reference ranges presented here represent the 0.25 and 97.5 fractiles derived after outliers were removed. Analytes in Tables III.C.1. to 3. indicated by # are those for which highly significant differences were found between women who did or did not use oral contraceptives and for which the reference ranges shown are derived only from non-oral contraceptive users.

REFERENCE

1. **Frohlich, J. J.,** Laboratory diagnosis of hyperlipidemia, in *Primary Hyperlipoproteinemias,* Steiner, G. and Shafrir, E., Eds., McGraw-Hill, New York, 1991, 143.

III.C.1. BIOCHEMISTRY REFERENCE RANGES

Table III.C.1. Biochemistry reference ranges for non-pregnant women aged 20 - 49 years (0.25 and 97.5 fractiles) #

Analyte	Units	Mean (SEM)	Reference Range
α-1-antitrypsin #	g/L	2.15 (0.04)	1.35 - 3.10
α-tocopherol	μmol/L	24 (0.5)	16 - 37
α-tocopherol:cholesterol ratio		5.2 (0.1)	3.8- 7.1
Alanine aminotransferase	U/L	18 (0.5)	5 - 29
Albumin #	g/L	41 (0.2)	36 - 46
Alkaline phosphatase	U/L	62 (1)	38 - 98
Amylase	U/L	50 (1)	27 - 82
Anion gap ((Na+K)-(Cl-HCO_3))	mmol/L	13 (0.15)	10 - 17
Aspartate aminotransferase	U/L	20 (0.05)	9 - 32
Bilirubin (conjugated)	μmol/L	<2	0 - 2
Bilirubin (unconjugated)	μmol/L	9 (0.4)	4 - 20
Calcium (total)	mmol/L	2.32 (0.01)	2.17 - 2.43
Calcium (ionized)	mmol/L	1.24 (0.01)	1.10 - 1.36
Ceruloplasmin #	mg/L	252 (4)	198 - 340
Chloride	mmol/L	105 (0.2)	101 - 108
Cholesterol (LDL) *	mmol/L	2.62 (0.08)	1.27 - 4.64
Cholesterol (HDL) *	mmol/L	1.41 (0.02)	0.87 - 1.99
Cholesterol (total) *	mmol/L	4.48 (0.08)	3.13 - 6.87
Colloid osmotic pressure	cm H_20	27 (0.6)	22 - 33
Complement C3 #	g/L	0.71 (0.01)	0.47 - 0.92
Complement C4	g/L	0.25 (0.01)	0.14 - 0.41
Copper #	μmol/L	15.4 (0.2)	11.7 - 20.5
Cortisol #	nmol/L	378 (13)	110 - 580

III.C.1. BIOCHEMISTRY REFERENCE RANGES (continued)

Table III.C.1. Biochemistry reference ranges for non-pregnant women aged 20 - 49 years (0.25 and 97.5 fractiles) continued #

Analyte	Units	Mean (SEM)	Reference Range
Creatine kinase	U/L	63 (3)	20 - 137
Creatinine	µmol/L	73 (1)	50 - 90
Ferritin *	µg/L	27 (2)	4 - 63
Gamma-glutamyl transferase	U/L	21 (0.4)	14 - 33
Glucose	mmol/L	4.7 (0.03)	3.9 - 5.5
Glutathione peroxidase	U/L	992 (33)	670 - 1312
Immunoglobulin A	g/L	2.22 (0.08)	0.68 - 4.08
Immunoglobulin G	g/L	13.20 (0.21)	8.93 - 18.21
Immunoglobulin M	g/L	1.94 (0.07)	0.69 - 3.51
Iron # *	µmol/L	17 (1)	6 - 32
Iron-binding capacity #	µmol/L	61 (1)	45 - 76
Lactate dehydrogenase	U/L	392 (6)	250 - 525
Magnesium	mmol/L	0.79 (0.01)	0.67 - 0.92
Osmolality	mosmol/kg	284 (5)	277 - 289
Phosphate	mmol/L	1.08 (0.01)	0.82 - 1.37
Potassium	mmol/L	4.3 (0.03)	3.7 - 4.8
Prealbumin #	mg/L	246 (5)	161 - 342
Retinol #	µmol/L	1.40 (0.03)	0.85 - 2.31
Retinol-binding protein #	mg/L	38 (1)	22 - 57
Selenium #	µmol/L	1.71 (0.02)	1.38 - 2.29
Sodium	mmol/L	139 (0.15)	136 - 142
Total CO_2 (bicarbonate)	mmol/L	26 (0.2)	22 - 29
Total protein	g/L	71 (0.4)	63 - 78

III.C.1. BIOCHEMISTRY REFERENCE RANGES (continued)

Table III.C.1. Biochemistry reference ranges for non-pregnant women aged 20 - 49 years (0.25 and 97.5 fractiles) continued #

Analyte	Units	Mean (SEM)	Reference Range
Thyroxine (total) #	nmol/L	104 (2)	77 - 141
Thyroxine (free)	pmol/L	13 (0.2)	9.5 - 16.5
Thyroxine-binding globulin #	mg/L	22 (0.4)	16 - 28
Thyroid-stimulating hormone	mU/L	1.8 (0.1)	0.4 - 4.6
Triiodothyronine #	nmol/L	1.9 (0.02)	1.3 - 2.4
Transferrin (TRF) #	g/L	2.85 (0.04)	2.18 - 3.89
TRF saturation		0.25 (0.02)	0.08 - 0.49
Triglyceride #	mmol/L	0.88 (0.04)	0.49 - 1.86
Urea	mmol/L	4.3 (0.1)	2.5 - 6.5
Uric Acid	µmol/L	238 (5)	154 - 325
Zinc	µmol/L	13.0 (0.14)	10.0 - 16.5
Zinc protoporphyrin	µmol/mol heme	43 (1)	26 - 64
Lead *	µmol/L	0.15 (0.01)	0.06 - 0.31

Reference ranges are derived from 2.5 and 97.5 fractiles of data from non-pregnant women 20-49 years. For variables where significant differences were found between women who were or were not using oral contraceptives, the ranges are derived only from the group not using oral contraceptives, and indicated by #.

* References ranges provided are those found in the healthy control population studied. They do not necessarily imply optimal levels for health.

III.C.2. HEMATOLOGY TESTS

Table III.C.2. Hematology reference ranges for non-pregnant women aged 20 - 49 years

Analyte	Units	Mean SEM)	Range
Red Cell Count	$X\ 10^{12}/L$	4.33 (0.03)	3.72 - 4.98
Hemoglobin	g/L	131 (0.8)	115 - 147
Hematocrit		0.387 (0.002)	0.343 - 0.426
MCV	fL	90.3 (0.32)	83.3 - 96.8
MCH	pg	30.6 (0.15)	27.4 - 33.8
MCHC	g/L	339 (0.83)	314 - 360
RDW		0.129 (0.0007)	0.116 - 0.155
WBC	$X\ 10^{9}/L$	5.5 (0.12)	3.1 - 8.7
Lymphocytes (auto)	$X\ 10^{9}/L$	1.8 (0.04)	1.1 - 2.7
Monocytes (auto)	$X\ 10^{9}/L$	0.2 (0.01)	0.1 - 0.5
Granulocytes (auto)	$X\ 10^{9}/L$	3.3 (0.09)	1.5 - 6.0
Neutrophils (manual)	$X\ 10^{9}/L$	3.16 (0.10)	0.92 - 5.40
Band Cells (manual)	$X\ 10^{9}/L$	66/120 had no band cells recorded. Range when band cells present was 0.001 - 0.341	
Platelet Count	$X\ 10^{9}/L$	259 (5)	166 - 381
Mean Platelet Volume	fL	10.4 (0.1)	8.0 - 12.4
PT	s	10.6 (0.06)	9.4 - 12.0
APTT	s	31.0 (0.23)	26.6 - 35.5
Fibrinogen	g/L	2.75 (0.05)	1.76 - 3.64

Part IV

Study Protocols, Materials, and Methods

IV.A. Recruitment of Subjects

Since our objective was to define physiological variation in blood indices during optimal pregnancy conditions, we selected our pregnant and control groups by *direct* sampling from a population characterized by predefined characteristics. For both groups, these were mainly well educated, professional, and technical hospital and university personnel, who were expected to be reasonably knowledgable about health and nutrition and had easy access to adequate food and health care. The only *a priori* exclusion criteria for controls were acute or chronic illness, alcoholism, or drug use.

Our pregnant subjects were recruited by their physicians from women in whom a low-risk uncomplicated pregnancy was anticipated. Exclusion took place *a posteriori* if on completion of the study it was observed that she had failed to meet the criteria of healthy uncomplicated pregnancy or labor. Since recruitment of controls was by poster advertisements in the hospital and of pregnant women by their physicians, the sample was not randomly selected but represented a motivated compliant group of individuals.

ETHICAL APPROVAL AND INFORMED CONSENT

The study protocols were approved by the Research Advisory Committee of Children's Hospital, the Research Advisory Committee of Grace Hospital, and the University of British Columbia Clinical Screening Committee for Research Involving Human Subjects. Informed consent was obtained from every participant prior to enrollment.

IV.B. Specimen Collection and Processing

SPECIMEN COLLECTION AND HANDLING PROCEDURES

To minimize preanalytical variation, specimen collection and processing protocols were rigidly adhered to. With the subject seated, a 25-mL blood specimen was collected by venipuncture, using minimal hemostasis. The following order of specimen collection was maintained throughout the study. For chemistry analyses, we drew blood into two siliconized Vacutainer tubes (Becton-Dickinson, Mississauga, Ontario) with no anticoagulant and one lithium-heparin Vacutainer. Both types of Vacutainer were suitable for trace element analysis. Citrate and EDTA-containing Vacutainer tubes were subsequently used for the coagulation and hematological specimens.

During collection and specimen processing, strict precautions were observed to prevent possible trace element contamination. After collection, the heparinized Vacutainer was protected from light by tin foil and placed in a container of ice. All tubes were kept upright to minimize contact of specimen with the Vacutainer stopper and transported within minutes of collection to the biochemistry and hematology laboratories.

Fibrinogen, PT, and PTT were measured in citrated plasma within 2 hours of collection. The automated blood panel was analyzed within 4 hours using the EDTA specimen. Aliquots of well-mixed whole blood were taken from the heparinized Vacutainer for zinc protoporphyrin and lead analyses. The remaining chemistry specimens were centrifuged at 0 to 4°C (2600 rpm, 10 min). Serum aliquoted for the Ektachem-700 chemistry panel was analyzed within 3 hours. The remaining serum and plasma was aliquoted and stored at 0 to 4°C or −70°C until used. Buffy coat was removed from the heparinized blood, red cells washed three times with cold isotonic saline, and packed red cell aliquots frozen at −70°C. Protein and trace element analyses were completed within 1 to 4 weeks.

IV.C. Instrumentation and Methods

EQUIPMENT AND METHODS

- Twenty-four analytes were measured on the Ektachem-700 (Eastman-Kodak, Rochester, NY): sodium, potassium, chloride, bicarbonate, urea, creatinine, uric acid, total calcium, phosphate, magnesium, glucose, bilirubin, alanine aminotransferase, alkaline phosphatase, amylase, aspartate aminotransferase, τ-glutamyl transferase, creatine kinase, lactate dehydrogenase, total protein, albumin, total and HDL cholesterol, and triglycerides.
- Ten proteins were measured on the Behring Nephelometer Analyzer (Behring Canada, Montreal, QU): α_1-antitrypsin, ceruloplasmin, complement fractions, C3 and C4, immunoglobulins IgA, IgG, and IgM, prealbumin, retinol-binding protein, and transferrin.
- Zinc, copper, selenium, and lead were measured on Varian atomic absorption (AA) spectrophotometers (Varian Canada, Georgetown, ON). Zinc was measured by flame AA, copper, selenium, and lead by electrothermal AA. Method information can be provided on request.
- Iron and iron-binding capacity were measured on the RA-1000 (Technicon Instruments, Tarrytown, NY) and on the Ektachem-700.
- Erythrocyte enzymes for assessment of thiamine, riboflavin, and pyridoxine status were measured on the Cobas Fara (Hoffman-La Roche, Ltd., Etobicoke, ON).* Method information can be provided on request.
- Ferritin was measured on the Quantum and IMx analyzers, α-fetoprotein on the IMx (Abbott Laboratories, Mississauga, ON).
- Thyroxine, cortisol, and related hormones were measured by radioimmunoassay and immunofluorescence on the Delfia fluorometer (Pharmacia, Baie d'Urfe, QU).
- The automated blood panel was measured on the S plus IV (Coulter, Burlington, ON).
- Coagulation studies were performed on the Technika Coagamate (Organon Technika, Scarborough, ON) and the ACL (Instrumentation Laboratories, Lexington, MA).

Lead and the erythrocyte enzymes were measured only at 12 weeks, term, and 6 weeks postpartum. All other variables were measured in accordance with the sampling schedule above.

* Retinol and tocopherol were measured by reverse phase HPLC (Waters Scientific, Mississauga, Ontario)

QUALITY ASSURANCE AND PROFICIENCY TESTING

All the tests evaluated were those in routine use in the Clinical Biochemistry and Hematopathology Laboratories at Children's Hospital. Throughout the period of the study, all these tests were subject to stringent ongoing internal quality control monitoring. In addition, data from laboratory performance on external quality control or proficiency programs including Wellcome, CAP, and special programs for trace metals and vitamins were carefully monitored to ensure consistent performance of our test procedures. The specific programs subscribed to include

- Wellcome Clinical Chemistry Quality Assessment Programme
- Wellcome Immunoassay Quality Assessment Programme
- CDC Lead Proficiency Program
- Quebec Toxicology
- EQUAS, The Foundation for Blood Research Lipid Plus Canadian Cholesterol Reference Foundation

DATA ANALYSIS

Computer analyses were carried out using the 1988 release of BMDP-PC Statistical Software.[1] The data were initially scanned for obvious outliers using histograms and scattergrams.[2,3] An outlier was a result which differed from the next result by a value greater than one third of the total range of results.[3,4] If no error in transcription was found, no clinical reason for the extreme value was determined, and the value was clearly an outlier, the value was excluded from the data set.[2,3] Data were analyzed by Student's t test, analysis of variance for repeated measures protocol, regression, and multivariate analysis.

REFERENCES

1. **Dixon, W. J.,** *BMDP Statistical Software Manual,* Dixon, W. J., Ed., University of California Press, Los Angeles, 1988.
2. **Solberg, H. E. and Grasbeck, R.,** Reference values, *Adv. Clin. Chem.,* 27, 1, 1989.
3. **Solberg, H. E.,** Statistical treatment of collected reference values and determination of reference limits, in *Reference Values in Laboratory Medicine,* Grasbeck, R. and Alstrom, T., Eds., John Wiley & Sons, Chichester, 1981, 193.
4. **Reed, A. H., Henry, R. J., and Mason, W. B.,** Influence of statistical method used on the resulting estimate of normal range, *Clin. Chem.,* 17, 275, 1971.

Index

INDEX

A

B

C

D

N

O

P

Q

R

S

W

Z